A Guide to the Electrical Contractor's Exam

A Guide to
the Electrical
Contractor's Exam

Alan W. Stanfield, M.B.A.

THOMSON

DELMAR LEARNING

Australia Canada Mexico Singapore Spain United Kingdom United States

A Guide to the Electrical Contractor's Exam
Alan W. Stanfield

Vice President, Technology and Trades ABU:
David Garza

Vice President, Technology Professional Business Unit:
Gregory L. Clayton

Director of Learning Solutions:
Sandy Clark

Managing Editor:
Larry Main

Product Development Manager:
Ed Francis

Product Manager:
Stephanie Kelly

Marketing Director:
Beth A. Lutz

Marketing Manager:
Taryn Zlatin

Marketing Specialist:
Marissa Maiella

Director of Production:
Patty Stephan

Production Manager:
Andrew Crouth

Content Project Manager:
Christopher Chein

Art Director:
Benj Gleeksman

Production Technology Analyst:
Thomas Stover

Editorial Assistant
Nobina Chakraborti

Library of Congress Cataloging-in-Publication Data
Stanfield, Alan.
  A guide to the electrical contractor's exam / Alan Stanfield.
      p. cm.
  Includes index.
  ISBN-13: 978-1-4180-6410-5
  ISBN-10: 1-4180-6410-6
  1. Electric engineering—United States—Examinations, questions, etc. I. Title.
  TK435.S73 2007
  621.319'2076—dc22

                                      2007028335

NOTICE TO THE READER

Publisher does not warrant or guarantee any of the products described herein or perform any independent analysis in connection with any of the product information contained herein. Publisher does not assume, and expressly disclaims, any obligation to obtain and include information other than that provided to it by the manufacturer.

The reader is expressly warned to consider and adopt all safety precautions that might be indicated by the activities herein and to avoid all potential hazards. By following the instructions contained herein, the reader willingly assumes all risks in connection with such instructions.

The publisher makes no representation or warranties of any kind, including but not limited to, the warranties of fitness for particular purpose or merchantability, nor are any such representations implied with respect to the material set forth herein, and the publisher takes no responsibility with respect to such material. The publisher shall not be liable for any special, consequential, or exemplary damages resulting, in whole or part, from the readers' use of, or reliance upon, this material.

Contents

Preface

An Overview

The purpose of this book is to provide candidates for the electrical contractor's license a study tool in their preparation for the exam. Most states require that an electrician be licensed in order to install electrical service. This means that an examination must be passed in order to obtain a State license. Most of the exam contains both electrical sections as well as business law sections. This book is divided into several chapters and covers both the electrical portion as well as the business law portion.

Readers of this book will be provided some reference material and explanations of certain topics. The book will work in conjunction with the *National Electrical Code*® and the Employer's Tax Guide. Numerous example problems will be performed step by step with a full explanation of each component of the problem. Multiple choice questions will then follow each subject with the answers provided in the back of the text. Finally, the reader will have the opportunity to demonstrate their knowledge with practice exams that closely imitate actual exam conditions. A complete answer key is provided in the appendix with detailed explanations of each solution.

This book attempts to simulate questions that resemble the test questions encounter on a state exam. Much of the book is dedicated to performing calculation type questions, however, most exams contain both calculation questions and short answer type questions. The overall objective of this book is to demonstrate how the National Electrical Code® pertains to an electrical exam and how to properly use the National Electrical Code® in an exam setting. The final chapter of the book provides an opportunity for the reader to practice with multiple choice questions which would simulate most exams.

How to Use This Book

The key to passing any exam or test is to be well prepared. Most people are not blessed with a photographic memory so there is some work involved in being successful on an exam. This book is laid out in such a way that it allows the reader to digest the material one section at a time. The best way to utilize this book is to develop a good knowledge of one chapter before moving on to the next since some of the information is sequential. You will notice that each chapter begins with a brief overview of what is covered then followed by the chapter objectives. You will also notice that most of the chapters contain Key Codes. These Key Codes are helpful in quickly locating what sections of the *NEC*® will be used within the specific chapter. This book is designed to work in conjunction with the *NEC*® so have a copy handy when you begin to use this book. Also, be sure to have a reliable calculator on hand for the calculation portions as well as pencils, pens, and highlighters. Find a quiet place to study where there will be few or no distractions. Study aids like this book are the most beneficial when you take your time preparing and not trying to cram in too much information in just a few days. In order to get the most out of this book, view it as not only as readable information but also as an interactive workbook.

Note to the reader

Note that not all state electrical license exams are the same and may cover a wide range of material. This book attempts to make the reader aware of some of the most popular material that may be covered on any exam pertaining to the National Electrical Code©. In order to insure accuracy, the reader should obtain all the available information that pertains to the state in which the exam is being taken.

Disclaimer

Every attempt has been made to ensure that the information and calculations in this book are correct, however, errors can occur. The author of this material does not assume responsibility for errors or omissions and is not liable for the use or misuse of the material contained in this document. Business law and tax information change annually so the reader should obtain the most recent information available in his or her final preparation for the exam.

National Electrical Code and NEC are registered trademarks of the National Fire Protection Association, Quincy, MA.

About the Author

Alan Stanfield is a native of Georgia and grew up in the home of an electrician. His Father, John retired as an electrician for the City of Griffin after nearly forty years of service. Alan graduated from Griffin Technical College with a Diploma in Residential & Commercial Wiring and then went on to pursue a Bachelor of Science degree in business from Lee University. There he worked as a maintenance electrician to help pay for college expenses. After graduation, he worked for an electrical service company as an electrical service technician. In 2001, Alan was hired as the industrial electrical program coordinator and instructor at Griffin Technical College. He continues to teach and write as an electrical instructor and operates his electrical contracting service. In addition, Alan holds a Master's of Business Administration degree from Mercer University.

Acknowledgements

Special thanks to my wife Kellie, to my family and friends and all those that have encouraged me along the way.

The Author and Thomson Delmar Learning would like to thank the following reviewers for their invaluable contributions throughout the development of this text:

Thomas Collins
Master Electrician; Professor of Electrical Technology
Gateway Community and Technical College
Cincinnati, OH

Richard L. Tarabula, Jr.
Journeyman Inside Electrician
Williamsville, NY

Paul V. Westrom
Assistant Professor
New England Institute of Technology
Warwick, RI

CHAPTER 1

INTRODUCTION AND THE NATIONAL ELECTRICAL CODE

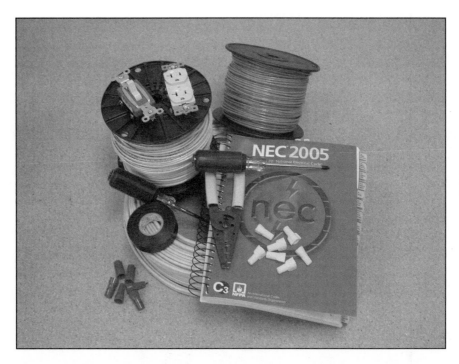

Chapter 1 covers some important elements of the electrical exam and the electrical license. It briefly describes the license itself and discusses some details about continuing education and state agreements. A math review will help sharpen your skills before sitting for the exam. Developing good math skills and an understanding of basic math fundamentals will prove to be vital on the exam. There are also some test-taking and study tips that will help narrow your focus throughout this book.

Objectives

- Explain the electrical contractor's license and testing formats
- Offer study tips and helpful hints
- Provide an overview of the National Electrical Code
- Review basic math skills
- Review general installation exam questions

The Electrical Contractor's License

Obtaining a state electrical contractor's license can be a very rewarding accomplishment. It will probably mean higher income and entrepreneurial opportunities as well. The license is an important step in becoming a confident and respected electrician. However, it is important to gain a thorough knowledge of the limitations of the license that you are pursuing. For example, some licenses may or may not qualify a person to pull permits, and other licenses may be restricted to certain voltages in some ways. Be sure to completely understand the requirements and the qualifying features of the license you desire.

One misconception about obtaining a license is that a person is immediately ready to start a business with little to no practical business experience. Passing the state exam does not solve all your problems; it is just the beginning of a fulfilling journey. If operating a business is not one of your strengths, you may consider working for a company as a licensed electrician. Some may argue that the headache of being your own boss is not always what it is cracked up to be, so be sure to consider all your options after you pass the test.

Introduction to the National Electrical Code

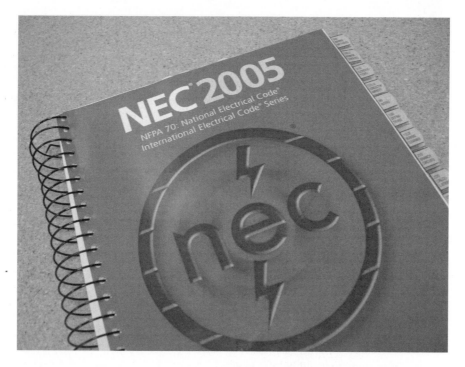

Being able to understand the different sections of the National Electrical Code ® (NEC ®) is imperative in achieving success on any state examination. The following paragraphs will describe each Article section, with a brief overview of the information it contains. It is not necessary to memorize each Article of the Code, but it is very important to become familiar with each Article in order to reduce the amount of time spent searching for the answers to exam questions.

The National Electrical Code is revised every three years (the last revision was dated 2005). Every three years different sections are subtracted or added, depending upon such factors as new proposals and changes in technology. The NEC is composed of nine chapters with a number of Articles in each chapter. Each chapter will be discussed in brief detail in an effort establish a general overview of what information is contained in each chapter.

Introduction

Article 90

The Code actually begins with Article 90, which covers the overall purpose of the Code as well as its scope. It helps the reader understand some basic elements of the Code and provides an introduction. Article 90 discusses where the NEC can or cannot be applied as well as the overall arrangement and enforcement of the Code.

Chapter 1

Articles 100–110

Article 100 provides some important definitions of key words used throughout the Code. It is a good place to define words that are essential in the Code and thereby clarify certain statements listed in the NEC. The Code can be confusing enough, but Article 100 can take some of the edge off all the confusion. Chapter 1 also includes items such as requirements for electrical installations like working space requirements and proper guarding from live electrical parts.

Chapter 2

Articles 200–285

This chapter covers requirements for identifying conductors, branch circuits, service calculations, overcurrent protection, and grounding. Chapter 2 is probably referred to more often than any other chapter in the NEC. Any exam taker should become very familiar with it because many exam questions can be asked from this chapter. Some states like to use this chapter because of all the calculation questions that can be asked. It is valuable to spend some time highlighting the important sections of Chapter 2.

Chapter 3

Articles 300–398

Chapter 3 is also a well-used section of the NEC because it too covers a lot of routinely referenced applications. This chapter covers wiring methods, burial requirements, all-important conductor ampacity, conduit and box fill, conductor applications, and electrical tubing applications. Table 310.16 is probably the most used table within the NEC and is one that will most certainly be referenced on any exam. As with Chapter 2, expect to spend quite a bit of time referring to Chapter 3 on exam day.

Chapter 4

Articles 400–490

Chapter 4 includes material such as switches, receptacles, panelboards, appliances, and transformers. However, Chapter 4 is best known for covering single- and three-phase motors. Many exams ask motor calculation questions due to their complexity, so become comfortable with Article 430 specifically. Out of all of the chapters in the NEC, these first four described so far are the most frequently used in the field and on examinations.

Chapter 5

Articles 500–590

Although the first four chapters from the NEC contain some of the most popular questions posed on the exam, expect to see some questions from the chapters that follow. Chapter 5 contains information about special locations such as health care facilities, hazardous locations, garages, gas stations, places of public assembly, and other special installations.

Chapter 6

Article 600–695

Similar to Chapter 5, Chapter 6 covers special equipment. Most exams do not rely heavily on Chapter 6 for exam questions, but any part of the NEC is fair game. This chapter contains information on special equipment like electric signs, cranes and elevators, electric welders, swimming pools, and solar electric systems. Probably the most referenced articles of this chapter would be welders, electric signs, and swimming pools.

Chapter 7

Article 700–780

Chapter 7 focuses on emergency systems, with most of the chapter's attention being paid to signal circuits and emergency backup systems. Don't expect many questions from this chapter, because much of the information covered is more involved with low-voltage applications.

Chapter 8

Articles 800– 830

Chapter 8 is one of the least used chapters for electricians in the NEC. It deals exclusively with communication systems such as radio and television systems. As with Chapter 7, don't expect many questions to be found in this chapter.

Chapter 9

Tables 1–12

Chapter 9 is the shortest chapter in the NEC but its importance should not be underestimated. Chapter 9 is different from the other chapters in that it contains only tables and notes. However, these tables are very important in performing calculations concerning conduit fill and voltage drop problems. Table 8 is particularly useful in determining conductor properties.

Annex .

The Annex is also an important part of the NEC. It contains information related to conduit fill similar to Chapter 9. All of the different types of conduits are displayed in the tables, which allows for quick reference. The Annex also contains example calculations that can be referred to on an exam.

Index

Finally, the Index provides the reader with necessary reference data to assist in locating key terms within the NEC. The Index is one section that you will utilize often during the exam as well as in the field. One of the most difficult aspects about using the NEC is simply finding what you are looking for. The Index can assist in locating some of those hard-to-find Code sections.

The Examination Method

Each state may use different types of exams that feature a variety of questions. Some exams are open book while others are closed book; yet others could allow a combination of the two. Some state exams feature business law and tax-related questions. There are also different types of licenses that can be achieved. For example, Georgia issues a restricted license (200 amps or less) and an unrestricted license (amperage unlimited). Other states may issue a journeyman's or a master's license.

The best way to prepare for a state electrical exam is to become familiar with the method and type of testing used in that state. The best way of doing that is to contact the state's licensing or construction board.

Continuing Education Units (CEUs)

Many states require that licensed electricians continue their education every year or two. Each state varies as to the method or the number of CEUs a license holder must obtain. It is the responsibility of the license holder to maintain these CEUs in order to keep the license active.

Reciprocal Agreements

There are states that allow license holders from neighboring or nearby states to conduct business across state lines. For example, a licensed electrician who obtained his or her license in Georgia can travel to Alabama and contract a job.

Seven Study Tips for Success

Before beginning any exam, there are some things that test takers need to know. For some, taking an exam is an easy task, but for most it can be an intimidating endeavor. As demonstrated below, there are helpful hints and tips that can alleviate some of the pressure of taking a long exam such as the electrical license test. Listed below are some tips that you may find useful in your journey towards the electrical license.

Tip 1: Start studying early.

You should begin studying for the exam several months in advance by working review problems and consulting the Code as much as possible. Becoming familiar with the Code will be your ticket to success on the exam. The key is not to memorize the Code book but to be able to identify the proper chapter and section to refer to when searching for an answer.

Tip 2: Look for key words.

When performing practice questions, identify key words or phrases that will lead you to the text in the Code. Oftentimes, if you read the question as a whole, it may be difficult to locate in the Code book. You may need to dissect the question and break it down into key words or terms.

Tip 3: Don't overstudy.

Stop every hour and take a break from studying. It is a fact that shorter and more consistent study sessions are more productive than long, tiring sessions. It is important to become familiar with the things that you need to know and not to waste time learning the things that you don't need to know. This is what sometimes happens when people overstudy for a test. Study guides like this one will help you narrow your study materials into more manageable sections.

Tip 4: Get some rest.

The night before the exam, get plenty of rest. The exam takes about eight hours to complete and most people need all the time they can get. If you are tired from lack of sleep, you may certainly see it in your results. Long tests like the state exam require that you remain alert and finish strong.

Tip 5: Get there early and be prepared.

The morning of the exam, get up in plenty of time to eat a good breakfast. Also, allow yourself adequate time to travel to the testing center. Arriving 30 to 45 minutes early is not a bad idea. If you get there late, not only will you feel rushed but you may be shown little mercy by the testing facilitators. Be sure to take along two dependable calculators, being certain that they are of the type that does not beep or make any other disturbing sounds. Also, take along several good pencils with erasers. In addition, if the state exam that you are taking allows you to use additional references other than the NEC, then take full advantage.

Tip 6: Control your time.

Time management is vital to your success on the exam. Try to answer every question, but skip over the more difficult questions, especially the ones that require substantial time to complete. First, answer the questions that you know and then come back to the more difficult ones. By doing so, you may avoid a lot of frustration by building your confidence and answering the easier questions first. Another effective way to control your time is to use tabs in the NEC, which will make locating chapters and specific references much easier. This is a rather small investment to make but can be one of the most useful study and test-taking aids available.

Tip 7: Eliminate the obvious.

Most of the time, your first instinct on an answer is your best choice. You are not going to get every question correct. Use the process of elimination (POE) to isolate the best choice. Most test questions contain four to five answer choices and, routinely, one or two of the answers are obviously not the best choice. Eliminate those obviously wrong choices and try to isolate the remaining choices that stand out the most as a possible solution.

Math Review

In order to be successful on the electrical exam, you will need to be able to calculate some relatively simple mathematical equations. However, even though most of the problems will require basic math skills, it is still very easy to make mathematical mistakes. It would be in your best interest to brush up before the exam on some of these basic skills, such as converting fractions to decimals, doing basic division, finding area and circumference, and using the metric system. Some of these formulas are located at the end of this book and can serve as a useful reference.

Order of Operations

Sometimes you may encounter a math problem that contains more than one type of operation, such as multiplication or addition. Some problems may combine addition, subtraction, multiplication, division, or the use of parentheses. When solving a math problem that combines more than one operation, you must yield to what is called the order of operations. Look at the problem below for a better understanding.

$$(100 - 75) + 25 = \underline{\quad}$$

Using the order of operations, you would need to apply the rules:

1. Perform all operations located inside the parenthesis first.
2. Next, perform any operations that contain roots or exponents.
3. Then, solve any multiplication or division operations that the problem may contain, from left to right.
4. Finally, perform any addition or subtraction, from left to right.

Refer back to the problem:

$$(100 - 75) + 25 = \underline{\quad}$$

1. The first step in solving this problem is to perform the operation located inside the parenthesis, which would be $(100 - 75) = 25$. So, if you rewrite the problem, now it should look like this:

$$25 + 25 = \underline{\quad}$$

2. The next step is to see if there are any roots or exponents in the problem, and in this example there are not, so ignore this step for this problem.
3. Then, solve any multiplication or division that may be in the problem. Since there is no multiplication or division in this example you may skip this step as well.
4. Finally, perform the simple addition operation to get the final answer, which is:

$$25 + 25 = \mathbf{50}$$

Example $20 + (50 - 20) - 10 = \underline{\quad}$

Solution $20 + (50 - 20) - 10 = \underline{\quad}$
$20 + (30) - 10 = \underline{\quad}$
$20 + 30 - 10 = \mathbf{40}$

Example $40 \times 20 / (15 + 5) = \underline{\quad}$

Solution $40 \times 20 / (15 + 5) = \underline{\quad}$
$40 \times 20 / (20) = \underline{\quad}$
$40 \times 20 / 20 = \underline{\quad}$
$40 \times 1 = \mathbf{40}$

Adding and Subtracting Decimals

Adding and subtracting decimals is a fairly simple mathematical calculation but if you are not careful, simple math problems can yield some big mistakes on an exam. The use of decimals is very common in the electrical field, so gaining a good knowledge of how to deal with decimals is very important. Demonstrated are some example problems that show how to add and subtract decimals.

Adding Decimals

When adding decimals, you need to make sure that all the decimal points of the numbers being added are properly aligned. Then, simply add the numbers together as usual and place the decimal point in the answer so that it aligns with the decimal points in the equation. Look at the examples below.

$$
\begin{array}{r}
4.6124 \\
+3.7832 \\
\hline
\textbf{8.3956}
\end{array}
$$

$$
\begin{array}{r}
210.67 \\
+122.77 \\
\hline
\textbf{333.44}
\end{array}
$$

$$
\begin{array}{r}
.099 \\
+227.38 \\
\hline
\textbf{227.479}
\end{array}
$$

Subtracting Decimals

Subtracting decimals is similar to adding, with the exception that the larger number should be placed on top with the smaller number on the bottom. You should subtract as with any other number but the decimal point must align as with addition. Look at the examples below.

$$
\begin{array}{r}
15.546 \\
-10.762 \\
\hline
\textbf{4.784}
\end{array}
$$

$$
\begin{array}{r}
145.88 \\
-34.10 \\
\hline
\textbf{111.78}
\end{array}
$$

$$
\begin{array}{r}
1.8457 \\
-0.0012 \\
\hline
\textbf{1.8445}
\end{array}
$$

The Metric System

It is probably most important to develop an understanding of the metric system before continuing on with this book. Many of the questions asked on an exam will probably include terms such as kilo, micro, and milli. Therefore, Table 1-1 presents some of the metric system units and some examples of how to recognize and convert the metric system.

Table 1-1: Metric System Conversions

Metric Symbol	Prefix	Decimal unit	Exponent
k	Kilo	1,000	10^3
	Base	1	
μ	Micro	0.000,001	10^{-6}

You will probably encounter some problems that use symbols like **kW** (kilo Watt) or **mA** (milli Amp). It is vital to pay attention to the unit of measure that is being used in order to get the correct answer. The question may even state the given information using units like kW (kilowatt) but only provide possible answers that are in base units like watts. Therefore, you must be able to spot such problems and be able to convert if necessary. Look at the following rules and examples:

Basic rules: (1) When converting *to* a base unit, multiple by the decimal unit.

(2) When converting *from* a base unit, divide by the decimal unit.

Converting to a base unit:	How many watts are in 1 kW?
	$1,000 \text{ watts} = 1 \text{ kW}$ so, $1 \text{ kW} \times 1,000 = \textbf{1,000 watts}$
Converting from a base unit:	Convert 20,000 watts into kW.
	$1 \text{ kW} = 1,000 \text{ W}$ so, $20,000 \text{ watts} / 1,000 = \textbf{20 kW}$
Converting from a base unit:	How many milliamps are in 10 amps?
	$1 \text{ amp} = 1,000 \text{ milliamps}$ so, $10 \text{ amps} / .001 = \textbf{10,000 mA}$
Converting from a base unit:	Convert 0.01 amps to milliamps.
	$0.01 / 0.001 = \textbf{10 mA}$
Converting to a base unit:	Convert 635 mA to amps.
	$635 \times 0.001 = \textbf{0.635 amps}$

Metric Conversion Practice Problems

Convert the following values:

1. 3/10 amps = _____ mA
2. 0.000285 amps = _____ μA
3. 45 ohms = _____ kilo ohms
4. 670 kilo ohms = _____ ohms
5. 230 millivolts = _____ volts
6. 2.78 mA = _____ amps

Area Calculations

When calculating area, you should consider the two-dimensional surface area. This type of calculation is common for figuring square footage of a home or building. It could also have other applications such as calculating yards of carpet or other floor coverings, for example. Area is usually expressed in units such as square feet, square yards, or square inches. The formula for calculating area is as follows (see Figure 1-1):

$$A(area) = W(width) \times H(height) \text{ or } L(length)$$

Figure 1-1

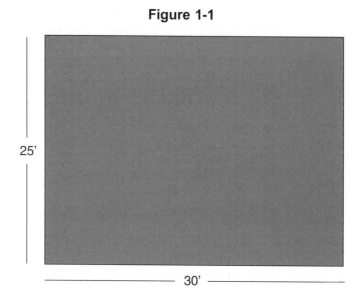

25'

30'

Example 1

What is the area of a room that measures 20' by 15'?

$$A = W \times L$$
$$A = 20' \times 15'$$
$$A = \textbf{300 square feet}$$

Example 2

What is the area of a home that is 60' long and 45' wide?

$$A = W \times L$$
$$A = 60' \times 45'$$
$$A = \textbf{2,700 square feet}$$

Example 3

What is the total square footage for a two-story home that has a first floor measuring 45' × 60' and a second floor measuring 30' × 30'?

$$A = W \times L$$
$$\text{1st floor area} = 40' \times 60'$$
$$= 2{,}700 \text{ sq ft}$$

$$2^{nd} \text{ floor area} = 30' \times 30'$$
$$= 900 \text{ sq ft}$$
$$1^{st} \text{ floor } (2{,}700) + 2^{nd} \text{ floor } (900) = \mathbf{3{,}600 \text{ total sq ft}}$$

Volume Calculations

Calculating volume is a little more complex in that it encompasses three dimensions. In order to figure the volume of an object, you must use not only the width and the height, but also the depth. Volume calculations are useful when figuring the amount of concrete needed for a slab or how much fill dirt is needed to backfill around a new home being constructed. The formula for volume is as follows (see Figure 1-2):

$$V(\text{volume}) = W(\text{width}) \times H(\text{height}) \times D(\text{depth})$$

Figure 1-2

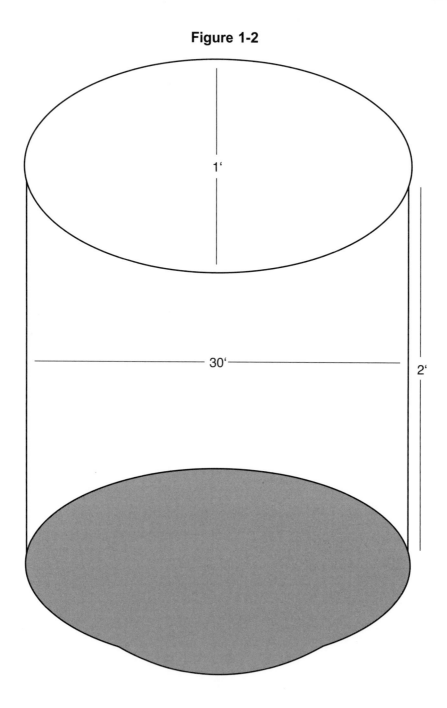

Example 1

A footing is being poured for a home. How much concrete is needed for a footing 30' by 2' by 1'?

$$V = W \times H \times D$$
$$V = 30' \times 2' \times 1'$$
$$V = \textbf{60 cubic feet}$$

Example 2

How much concrete is needed for a slab that is 40' by 20' and has a depth of 6"?

$$V = W \times H \times D$$

When calculating this problem, you must be careful to convert the 6" to a decimal. If you are not careful, punching in a 6 for the depth would cause a serious miscalculation.

$V = 40' \times 20' \times .5'$

$V = $ **400 cubic feet**

Math Practice Problems

Answer the following math problems:

1. $20 \times 53 = $ _____

2. $16 \times 235 = $ _____

3. $5,765 \times 18 = $ _____

4. $1,260 / 30 = $ _____

5. $34,000 / 16 = $ _____

6. $1,200 / .05 = $ _____

7. What is the square footage for a room that measures 12' by 10'?
   a. 144
   b. 120
   c. 1,200
   d. not listed

8. A certain container is 24" wide, 12" high, and 6" deep. What is the volume of the container?
   a. 1,450 cu in
   b. 1,450 cu ft
   c. 1,728 cu in
   d. 1,728 cu ft

9. What is the square footage of a home that measures 70' long and 45' wide?
   a. 2,200
   b. 3,000
   c. 3,150
   d. not listed

10. A concrete pad is to be poured for an outdoor light fixture. It should measure 2' × 2' × 1.5'. How many cubic feet of concrete are needed?
    a. 1 cu ft
    b. 2 cu ft
    c. 4 cu ft
    d. 6 cu ft

General Installation

Many of the questions on the exam will be of the general electrical installation type. A number of these questions will come from Articles 90, 100, and 110 of the NEC; however; they could also come from any other section of the NEC as well. They may be questions that cover definitions and terms, conductor identification, and other types of questions that can be found within different articles of the NEC. General questions are those that usually do not require any calculations to solve. They normally just require the test taker to locate the proper Code reference and select the best possible solution provided. General installation questions really test your knowledge and ability to use the NEC in order to find the correct response. This unit will demonstrate these types of questions and provide some examples for practice.

Example 1

The Code does not cover which of the following installations of electrical conductors?

    A. carnivals

    B. floating buildings

    C. mobile homes

    D. ships

  Answer: **D, ships**   90.2 (B)(1)

Example 2

A conductor encased with a material of composition that is not recognized by the Code is:

    A. conductor, insulated

    B. conductor, covered

    C. conductor, bare

    D. none

  Answer: **B, conductor, covered**   Article 100 page 27

Example 3

In order for a piece of equipment to be considered in sight by definition, it must be located within how many feet of the equipment?

    A. 50

    B. 40

    C. 60

    D. 15

  Answer: **A, 50**   Article 100 page 29

Example 4

A bored hole in a wooden framing member must not be made less than _____ inches from the nearest edge of the wooden member. (See Figure 1-3)

    A. 1

    B. 1¼

    C. 1½

    D. 2

  Answer: **B, 1¼**   300.4(A)(1)

Figure 1-3

The next example demonstrates a common question that is asked on many exams. This is an example of how closely related the answers can be and how an exam taker needs to pay close attention to every detail of the question and answers. This is a question that may appear to be fairly simple to some but is often missed because the reader does not fully examine all of the answer choices. Refer to Figure 1-4 for a common field application.

Example 5

According to Article 440.14, the disconnecting means for an air-conditioning unit must be located _____ . (See Figure 1-4)

    A. on panels designed for access
    B. within sight of the unit
    C. within sight of the unit and readily accessible
    D. only readily accessible

Answer: **C, within sight of the unit and readily accessible**   440.14

Figure 1-4

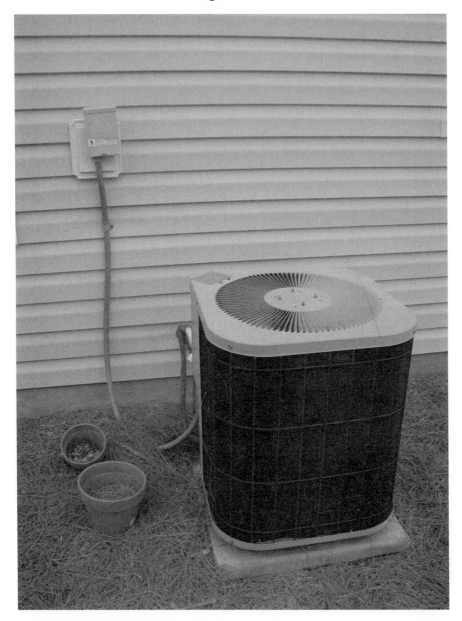

General Installation Practice Questions

1. What section of the NEC covers the required number of small appliance branch circuits to be installed?
   a. 210.11(C)(1)
   b. 210.11(A)
   c. 210.52(B)(1)
   d. none listed

2. What is the maximum height that an outdoor outlet can be installed above grade?
   a. 4'
   b. 5' 6"
   c. 6' 6"
   d. 7'

3. Heavy-duty lampholders are calculated at a minimum of _____ volt amps.
   a. 200
   b. 600
   c. 450
   d. 800

4. What section of the NEC prohibits vegetation to be used as a method of support of overhead service conductors?
   a. 231.10
   b. 240.15
   c. 230.29
   d. 230.28

5. Which of the following is not considered to be a standard size fuse?
   a. 125
   b. 350
   c. 45
   d. 75

6. A ground ring must be made from a conductor that is no smaller than _____ AWG.
   a. 2
   b. 6
   c. 4
   d. 1/0

7. Type NM cable is permitted to be used for the following purposes except:
   a. in cable trays
   b. in multifamily dwellings
   c. embedded in concrete
   d. in air voids of masonry block

8. Rigid metal conduit (1½") is required to be supported at a maximum distance of _____ feet.
   a. 12
   b. 20
   c. 16
   d. 14

Summary

After reading and working through this chapter, you should have a better understanding of how the exam will be set up and be more familiar with the National Electrical Code. You should also be better prepared for how to take the exam and what to expect when you arrive at the testing center. This chapter laid the groundwork for the future chapters that you will encounter by discussing basic math problems and by providing example problems that will help you decipher that NEC. In order to make the most out of Chapter 1, spend some time flipping through the NEC before moving on to Chapter 2 to grasp a better understanding of the overall layout of the NEC.

CHAPTER

2

ELECTRICAL THEORY

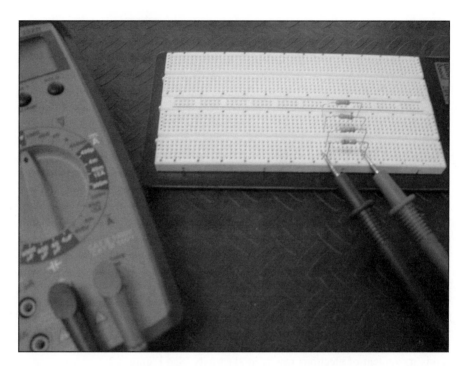

Chapter 2 addresses the most widely used set of formulas in the electrical industry, which is Ohm's Law. Ohm's Law forms the basis for most electrical calculations, and every electrician should have a good knowledge about it. This chapter also discusses atomic structure, along with series circuits, parallel circuits, and voltage drop calculations. Make sure that you have a solid understanding of Ohm's Law before moving on to the next chapter.

Objectives

- Explain atomic structure and how electricity works
- Describe the difference between AC and DC current
- Explain Ohm's Law and how to apply it
- Discuss series and parallel circuits
- Determine how to calculate voltage drop

Atomic Structure

In order to understand the theory of electricity, it is necessary to study the atom. The atomic structure of an atom is made up of protons, neutrons, and electrons. Protons have a positive charge, neutrons are neutral, and electrons have a negative charge. Protons and neutrons form the nucleus of the atom while the electrons orbit the nucleus. See Figure 2-1.

Figure 2-1

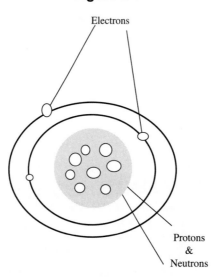

Electrons

Protons
&
Neutrons

As the electrons orbit around the nucleus, the electrons located on the outer shell or ring are called the valance electrons. The valance electrons are capable of moving from one atom to another. As the electrons spin around the nucleus, centrifugal force will cause these outermost electrons to spin out of orbit and strike another atom. This striking action will dislodge the electron from the second atom and the electron from the first will then orbit the atom it has struck. This action continues repeatedly and thus this movement of electrons results in electricity. See Figure 2-2.

Figure 2-2

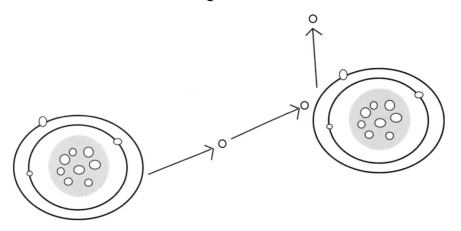

DC vs. AC: A Bit of History

In the late 1800s, there was a battle between direct current (DC) and alternating current (AC). Some believed that DC was the safest and most effective type of voltage available, while others argued that AC was the most efficient. Thomas Edison supported the argument that DC was the best way to light up the world, while a man by the name of Nikola Tesla suggested that AC was a much better way to meet the demands for the needed electricity. Some argued that AC power was too dangerous to operate, but the flexibility of AC eventually won out over DC. The score was ultimately settled at the 1904 World's Fair when two bids were submitted by two electric companies, one DC company that was led by Edison and one AC company that was lead by George Westinghouse. The Westinghouse bid won out by a landslide and the rest is history.

The main reason for the acceptance of AC over DC is that AC power can be transmitted over much greater distances due to the fact that it can be stepped up or down by the use of transformers. DC, on the other hand, is limited in the overall distance that it can be transmitted and would require many more substations in order to be as effective as AC power. See Figure 2-3 for an example of an AC sine wave.

Figure 2-3

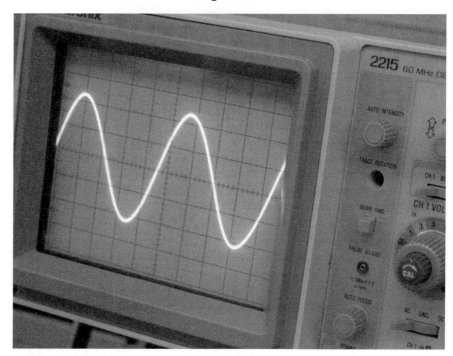

Ohm's Law

Ohm's Law is perhaps the most frequently used set of formulas in the electrical field. Developing a complete understanding of how to use and apply the correct Ohm's Law formula is necessary not only for the electrical exam but also for field applications as well. In this section of Chapter 2, you will see the importance of the math review provided in Chapter 1 in relation to Ohm's Law. Take a look at Ohm's Law in its simplest form, featured in Figure 2-4. All electricians, whether licensed or not, should be able to perform simple Ohm's Law calculations without even referring to an Ohm's Law chart or wheel.

Figure 2-4

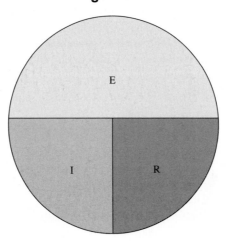

How to Use Ohm's Law

Referring to Figure 2-4, notice that there are three letters inside the circle and that the circle is divided into three different sections. The light shaded section contains the letter E, which represents voltage. The medium shaded section contains the letter I, which represents current. The dark shaded section contains the letter R, which represents resistance. Notice in Figure 2-4, how the circle has one horizontal line across the middle and half of a vertical line between I and R. The horizontal line represents division and the half vertical line represents multiplication. Therefore, if you are searching for I, you would divide E by R. If you are searching for E, you would multiply I × R, and so on.

In order to determine any one value, either E, I, or R, two other values must be known. For example, if voltage and resistance are known, you can find current by using Ohm's Law. Simply cover I, and you will find that calculating I means dividing E by R. See Figure 2-5.

Figure 2-5

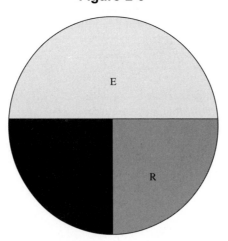

Refer to Figure 2-6. In this example, I and R would be the known values. Suppose that a question asked for you to determine how many volts are present in a circuit that contains 2 amps and 50 ohms of resistance. By blocking out E, you can see that I should be multiplied by R.

2 amps × 50 ohms = **100 volts**

Figure 2-6

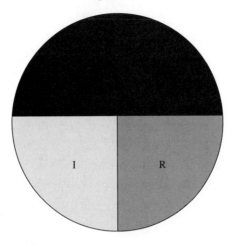

Figure 2-7 illustrates a more complex Ohm's Law wheel. However, the same concept applies. First, find what value you are searching for, whether it be I (current), E (voltage), R (resistance), or P (power). Then locate the two known values that pertain to the appropriate unknown value and solve.

Figure 2-7

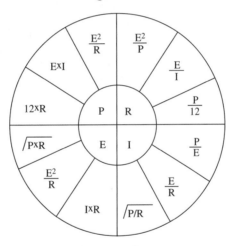

Ohm's Law formulas determine power, voltage, current, and resistance. They are used in a number of electrical applications. Notice Figure 2-8. Say, for example, a problem asks you to find voltage; you would first locate E on the wheel. Then, you would need to know what two values are given in the equation. Say that current and resistance are the given values. Finally, you would determine which of the three formulas that achieve E would apply. In this case, the formula I × R would be the correct choice.

Figure 2-8

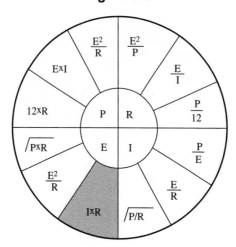

Here are some typical examples of the types of Ohm's Law questions that might be asked on an exam. Refer to the Ohm's Law wheel that follows each example in order to see how the problem is solved.

Example 1

A lamp operates on 120 volts and has a resistance of 90 ohms. What is the current draw of the lamp? Refer to Figure 2-9.

$$I = \frac{E}{R} \quad \frac{120 \text{ volts}}{90 \text{ ohms}} = \textbf{1.33 amps}$$

Figure 2-9

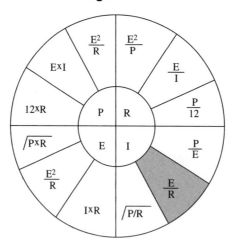

Example 2

A 240-volt heater draws a current of 5 amps. How much power is consumed by the heater? Refer to Figure 2-10

$$P = E \times I \quad 240 \times 5 = \textbf{1,200 watts}$$

Figure 2-10

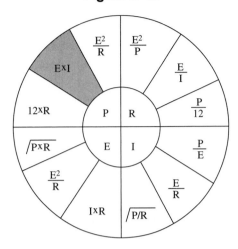

Ohm's Law Practice Problems

1. A 120-volt lamp has a resistance of 90 ohms. How much current is flowing through the lamp?
   a. 1.33 amps
   b. 2.25 amps
   c. 1.25 volts
   d. 4 amps

2. How much resistance does a lamp have if it is connected to 120 volts and has a current of 0.7 amps?
   a. 235 ohms
   b. 90 ohms
   c. 171 ohms
   d. 75 ohms

3. A 10,000-ohm resistor hás a meter-measured current of .05 amps. What is the voltage applied to the resistor?
   a. .35 volts
   b. 35 volts
   c. 500 volts
   d. none listed

4. How much current flows through a 120-volt, 250-watt bulb?
   a. 2.08 amps
   b. 0.24 amps
   c. 8.25 amps
   d. none listed

5. A dishwasher has a heating element that is rated at 3,000 watts and is connected to a 240-volt circuit. What is the current draw?
   a. 60 amps
   b. 0.06 amps
   c. 12.5 amps
   d. 20 amps

6. What is the power consumed by an electric heater that draws 4.5 amps when connected to a 120-volt outlet?
   a. 40 watts
   b. 540 watts
   c. 120 watts
   d. 180 watts

7. A 3-amp toaster, 2.5-amp coffee pot, and a 3.5-amp blender are all operating on one 120-volt circuit. What is the total amperage of the circuit if only these three appliances are used in the circuit?
   a. 6 amps
   b. 9 amps
   c. 12 amps
   d. not listed

8. Referring to question 7, what is the amount of resistance in the circuit?
   a. 13.33 ohms
   b. 1,080 ohms
   c. 75 ohms
   d. 1,200 ohms

9. How much current is needed to light a 400-watt light bulb at 277 volts?
   a. 2.64 amps
   b. 6 amps
   c. 1.44 amps
   d. 0.69 amps

10. A simple 12-volt circuit has a meter reading of 90 ohms. What is the amperage of the circuit?
    a. 133.33 mA
    b. 7.5 A
    c. 60 mA
    d. not listed

Series Circuits

Rules for series circuits:

1. Total resistance is equal to the sum of all resistors added together.
2. Current remains the same throughout the circuit.
3. Total voltage is equal to the voltage drops at each resistor.

For the following sections concerning series, parallel, and series-parallel circuits, apply the listed abbreviations to the problems:

RT = total resistance   ET = total voltage   IT = total current

R1, R2, R3... = resistor number

Figure 2-11

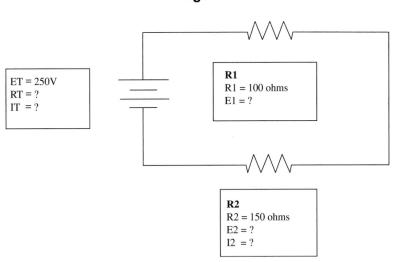

Example 1

Find the total resistance in the circuit Figure 2-11.

$$R1 + R2 = RT$$
$$100 \text{ ohms} + 150 \text{ ohms} = \textbf{250 ohms}$$

Example 2

What is the current measured at resistor 2?

$$E / R = I$$
$$250V / 250 \text{ ohms} = \textbf{1 amp}$$

Example 3

What is the voltage drop at resistor 1?
    Since we know that resistor 1 has a resistance of 100 ohms and a current of 1 amp, we can apply
    Ohm's Law. $I \times R = E$

$$1 \text{ A} \times 100 \text{ ohms} = \textbf{100 volts}$$

Example 4

What is the voltage drop at resistor 2?
    Since we know that current remains the same in a series circuit, we can apply Ohm's Law. $I \times R = E$

$$1 \text{ A} \times 150 \text{ ohms} = \textbf{150 V}$$

Series Practice Problems

Refer to the circuit in Figure 2-12 for questions 1 through 5.

Figure 2-12

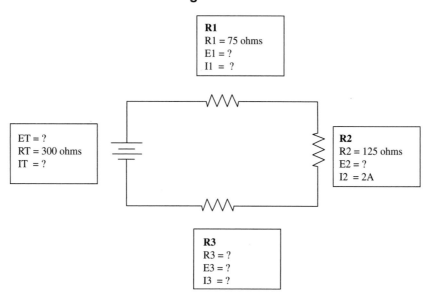

1. What is the total current (IT) in the circuit?
   a. 1 amp
   b. 2 amps
   c. 3 amps
   d. 4 amps

2. What is the voltage drop (E3) at resistor 2?
   a. 75 volts
   b. 250 volts
   c. 0 volts
   d. none listed

3. What is the resistance value (R3) of resistor 3?
   a. 50 ohms
   b. 125 ohms
   c. 200 ohms
   d. 100 ohms

4. The total voltage (ET) applied to the circuit is _____ volts.
   a. 600
   b. 200
   c. 400
   d. none listed

5. In order to solve any series circuit problem, what information must be available?
   a. RT must be given
   b. ET must be given
   c. at least one value at each resistor
   d. at least two values at one resistor or at the power supply

Parallel Circuits

Rules for parallel circuits:

1. Current changes at each resistor. The total current equals the sum of the branches.
2. Voltage remains the same throughout the circuit.
3. Total resistance is less than the least valued resistor. Total resistance must be calculated using one of the following formulas.

When you solve parallel circuits, it is important to try to solve for RT as early in your calculations as possible. There are three different methods with which to solve for RT. Each of these methods is discussed below.

Method #1
If all of the resistors are of the same value, then you can use this formula:

$$RT = R / N$$

R is the resistance value of one of the resistors and N is the number of resistors in the circuit. Refer to Figure 2-13.

Figure 2-13

R1   R2   R3   RT = ?

12 ohms   12 ohms   12 ohms

Example 1

Refer to Figure 2-13. What is the total resistance of the circuit?

$$RT = R / N \quad \text{or} \quad RT = 12 / 3 = \textbf{4 ohms}$$

Method #2
If the resistors are of different values, you must solve for RT using a different method than the previous one described. There are two methods to determine RT and the first is the product over the sum method:

$$RT = R1 \times R2 / (R1 + R2)$$

Resistor value 1 is multiplied by resistor value 2 and then divided by the sum of resistor value 1 and resistor value 2.

Figure 2-14

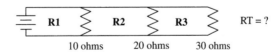

R1   R2   R3   RT = ?

10 ohms   20 ohms   30 ohms

Example 2

Refer to Figure 2-14. What is the total resistance of the circuit?

$$RT \text{ for } 1\&2 = 20 \times 30 / 20 + 30$$
$$= 600 / 50$$
$$= 12$$

Notice that 12 ohms is not the value we are looking for to find RT. R2 and R3 are now combined to form a new resistor with a value of 12 ohms. The circuit has been broken down from three resistors to two resistors. See Figure 2-15.

Figure 2-15

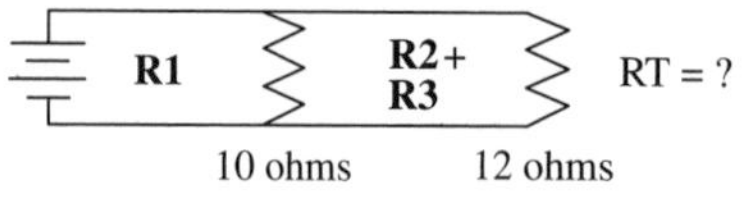

Now that we have broken down the problem to just two resistors, we can complete the problem by repeating the same formula except injecting the value of the combined resistors as the R2 value.

$$RT = 10 \times 12 / 10 + 12$$
$$= 120 / 22$$
$$= \textbf{5.45 ohms}$$

Method #3

The third method to solve for RT in a parallel circuit is to use the reciprocal formula.

$$RT = \frac{1}{1/(R1 + 1/R2 + 1/R3 + 1/Rn)}$$

This formula is the most useful in that no matter how many resistors are in a particular circuit, RT can be found in one formula.

Figure 2-16

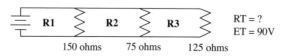

Example 3

Refer to Figure 2-16. What is the total resistance of the circuit?

$$RT = \frac{1}{1/100 + 1/200 + 1/150}$$
$$= \frac{1}{.01 + .005 + .0067}$$
$$= \frac{1}{.0217} = \textbf{46.08 ohms}$$

Parallel Circuit Practice Problems

Refer to Figure 2-17 below for questions 1 and 2.

Figure 2-17

1. What is the total resistance (RT) in the circuit?
   a. 200 ohms
   b. 85.71 ohms
   c. 125.14 ohms
   d. 75 ohms

2. What is the total voltage (ET) of the circuit?
   a. 90 volts
   b. 30 volts
   c. 270 volts
   d. 60 volts

Refer to Figure 2-18 for questions 3 through 5.

Figure 2-18

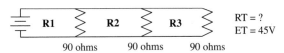

3. What is the total resistance (RT) of the circuit?
   a. 90 ohms
   b. 60 ohms
   c. 45 ohms
   d. 30 ohms

4. What is the voltage (E2) at R2?
   a. 45 volts
   b. 90 volts
   c. 30 volts
   d. 15 volts

5. What is the total current (IT) of the circuit?
   a. .67 amps
   b. 1.5 amps
   c. 2 amps
   d. 3.21 amps

Voltage Drop Calculations

Voltage drop occurs as a result of the resistance of a conductor. As the length of a conductor increases, the resistance also increases and therefore reduces the intensity of the voltage. Voltage drop should definitely be considered on long runs of cable such as with outdoor signs and long branch circuit runs. The NEC covers voltage drop in 210.19(A)(1) FPN4 and 215.2(A)(4) FPN. Also, refer to Table 8 of the NEC for the CM (circular mils) on page 635. Notice Figure 2-17 and refer to the notes below for more details about voltage drop.

- Code recommends but does not require a 3% drop or 3.6 volts for a 120 V circuit.

- Voltage drop $= \dfrac{2 \times K \times D \times I}{CM}$

- 2 = constant

- K = 12.9 for copper and 21.2 for aluminum

- D = distance from the source to the load

- CM = circular mils

When calculating for three-phase circuits, 1.732 must be used instead of 2 in this equation.

Example 1

What is the voltage drop on a 120 V, single-phase circuit using a #14 copper wire with a load of 10 A and a distance of 90 feet from the panel? See Figure 2-19.

$$\frac{2 \times 12.9 \times 90 \times 10}{4110} = \textbf{5.64 volts} \quad \textit{Exceeds Code limit of 3.6 V}$$

Figure 2-19

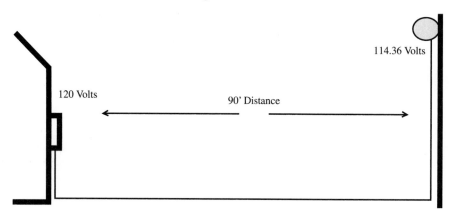

120 Volts

90' Distance

114.36 Volts

Voltage Drop Practice Problems

1. What is the voltage drop on a 120 V, single-phase circuit using a #10 copper wire with a load of 10 A and a distance of 90 feet from the panel?
   a. 10.0 V
   b. 12.4 V
   c. 4.5 V
   d. 2.24 V

2. What size copper wire is needed for a 15-amp load that is located 185 feet from a 120 V panel?
   a. #4
   b. #6
   c. #8
   d. #10

3. What is the voltage drop on a 120 V circuit with a #12 copper wire at a distance of 100 feet? The circuit supplies a 7-amp pump motor.
   a. 2.76 V
   b. 3.45 V
   c. 4.75 V
   d. none listed

4. According to Table 8, how many ohms per 1,000 feet does a #12 AWG solid copper wire have?
   a. 1.98
   b. 1.21
   c. 1.93
   d. 3.14

5. What is the voltage drop on a 240 V circuit with a #10 copper wire at a distance of 250 feet? The circuit supplies a 12-amp irrigation pump.
   a. 7.457 volts
   b. 6.25 volts
   c. 8 volts
   d. 5.45 volts

6. Referring to question 5, what is the voltage drop allowed by the NEC and does the circuit exceed the limit?
   a. 3.6, yes
   b. 7.2, no
   c. 9, no
   d. 7.2, yes

Summary

In this chapter, some very important topics have been covered. The importance of Ohm's Law cannot be overstressed. Ohm's Law can be found at the core of most electrical equations and it most certainly will be used on the electrical exam. Don't take electrical theory for granted, because it makes for some great exam questions. Many electrical workers don't have a good knowledge of theory and the exam writers are aware of that, so make sure you study hard on the theory section.

CHAPTER 3
BRANCH CIRCUIT AND CONDUCTOR AMPACITY

When installing branch circuits and conductors, take care to ensure that you have performed the proper calculations. It is necessary to take into account a number of variables in order to establish a safe and effective installation. The movement of electrons through a conductor produces heat, and if this heat is not allowed to dissipate properly, it can cause damage to both the conductor and the property.
Table 310.16 is commonly used to determine the allowable **ampacity** of a conductor; however, the table is based on a specific temperature of 30°C (86°F) and not more than three current-carrying conductors in a raceway. In most installations, these conditions are not always maintained, so

correction factors and derating must be calculated into the equation. For example, conductors run through an attic space could be exposed to **ambient temperature** in excess of 120°F. This type of condition would certainly decrease the ampacity of the conductors by not allowing for adequate heat dissipation.

Most exam questions concerning ampacity will state different ambient temperature conditions and possibly more than three current-carrying conductors in a raceway. A complete understanding of Table 310.16 and Table 310.15(B)(2)(a) is vital in order to ensure success on the exam.

Ampacity — The amount of current a conductor can carry without damage

Ambient temperature — The temperature around a conductor or equipment

Adjustment factors — When more than three conductors are installed in a raceway, the ampacity must be reduced

Objectives

- Discuss and calculate conductor ampacity

- Define ambient temperature and how it affects ampacity

- Determine derating factors for conductors

- Calculate **adjustments factors** for raceways containing more than three conductors

- Discuss branch circuit loading

Key Codes

Table 310.16	310.15(B4b) Neutral counted
310.15(B2a) Adj. factors	310.15(B4c) Nonlinear load neutrals
310.15(B2) ex. 1 Different systems	310.15(B5) Grounding/Bonding wires
310.15(B2) ex. 2 Cable trays	310.15(B2) ex. 4 Outdoor trenches
310.15(B2) ex. 3 Nipples (24")	310.15(B4a) Balanced neutrals

Table 3-1: Ampacity, Table 310.16

For this chapter, refer to Table 310.16 and Table 310.15(B)(2)(a). Other articles will also be used in determining branch circuit loading and conductor ampacity. These other articles will be discussed later in this chapter. Table 310.16 is probably the most misused and misinterpreted table in the NEC. The reason for such misinterpretation can be found by studying the title heading: *Not more than **three** conductors...temp. of 30C or 86F.*

Table 3-1: Table 310.16 Ampacity

Size AWG	60 C	75 C	90 C	60 C	75 C	90 C
	Copper			**Aluminum**		
	Types TW, UF	Types RHW, THHW, THW, THWN, XHHW, USE, ZW	Types TBS, SA, SIS, FEP, FEPB, MI, RHH, RHW-2, THHN, THHW, THW-2, THWN-2, USE-2, XHH, XHHW, XHHW-2, ZW-2	Types TW, UF	Types RHW, THHW, THW, THWN, XHHW, USE, ZW	Types TBS, SA, SIS, FEP, FEPB, MI, RHH, RHW-2, THHN, THHW, THW-2, THWN-2, USE-2, XHH, XHHW, XHHW-2, ZW-2
18	-	-	14	-	-	-
16	-	-	18	-	-	-
14*	20	20	25	-	-	-
12*	25	25	30	20	25	25
10*	30	35	40	25	30	35
8	40	50	55	30	40	45
6	55	65	75	40	50	60
4	70	85	95	55	65	75
3	85	100	110	65	75	85
2	95	115	130	75	90	100
1	110	130	150	85	100	115
1/0	125	150	170	100	120	135
2/0	145	175	195	115	135	150
3/0	165	200	225	130	155	175
4/0	195	230	260	150	180	205
250	215	255	290	170	205	230
300	240	285	320	190	230	255
350	260	310	350	210	250	280
400	280	335	380	225	270	305
500	320	380	430	260	310	350
600	355	420	475	285	340	385
700	385	460	520	310	375	420
750	400	475	535	320	385	435

(continued on next page)

Table 3-1 *(continued)*

Size AWG	60 C	75 C	90 C	60 C	75 C	90 C
	Copper			Aluminum		
	Types TW, UF	Types RHW, THHW, THW, THWN, XHHW, USE, ZW	Types TBS, SA, SIS, FEP, FEPB, MI, RHH, RHW-2, THHN, THHW, THW-2, THWN-2, USE-2, XHH, XHHW, XHHW-2, ZW-2	Types TW, UF	Types RHW, THHW, THW, THWN, XHHW, USE, ZW	Types TBS, SA, SIS, FEP, FEPB, MI, RHH, RHW-2, THHN, THHW, THW-2, THWN-2, USE-2, XHH, XHHW, XHHW-2, ZW-2
800	410	490	555	330	395	450
900	435	520	585	355	425	480
1000	455	545	615	375	445	500
1250	495	590	665	405	485	545
1500	520	625	705	435	520	585
1750	545	650	735	455	545	615
2000	560	665	750	470	560	630

[Reprinted with permission from the NFPA 70-2005, *National Electrical Code*®, Copyright © 2004, National Fire Protection Association, Quincy, MA 02269. This reprinted material is not the complete and official position of the NFPA on the referenced subject, which is represented only by the standard in its entirety.

Key Notes about Table 310.16

Table 310.16 is often misused because of one of the notes that is located at the bottom of the table, underneath the correction factor section. Notice in your code book, or in Table 3-1, that an asterisk (*) appears beside the AWG sizes 14, 12, and 10. The asterisk refers to 240.4(D), which limits a #14 to 15 amps, a #12 to 20 amps, and a #10 to 30 amps. Most exams ask questions that involve #14, #12, and #10 for this reason. If most people were asked how many amps a #12 THW conductor will carry, they would respond by saying, 35. However, that is not the case, because 240.4(D) limits a #12 to only 20 amps regardless of what Table 310.16 says. As you can see, this can be very tricky on an exam.

In addition to the asterisk used in the table, there is another restriction that can be found in the title of Table 310.16. Notice that is says, *Not More Than Three Current-Carrying Conductors in Raceway, Cable, or Earth*. If a conduit or raceway contains more than three current-carrying conductors, the ampacity of the conductors will be reduced. If more than three current-carrying conductors are present, you must use a correction factor located at the bottom of Table 310.16. An example of this situation will be demonstrated later in this chapter. So, when a question is asked on an exam about conductor ampacity, be sure to ask yourself two important questions: (1) Is the question referring to a #14, #12, or #10 conductor? (2) Are there more than three current-carrying conductors in the raceway?

One other note about Table 310.16 that you should be aware of is that this table does not include conductors that are installed in **free air**. Be sure to read the question very carefully when asked about items concerning ampacity. If the question refers to free air, notice that the answer would be calculated not according to Table 310.16 but rather Table 310.17. Notice that the heading for Table 310.17 includes the words "Free Air."

Free air — Conductors installed outdoors, for example, service entrance cables, usually allow for smaller conductors because of air circulation. See Figure 3-1.

Figure 3-1

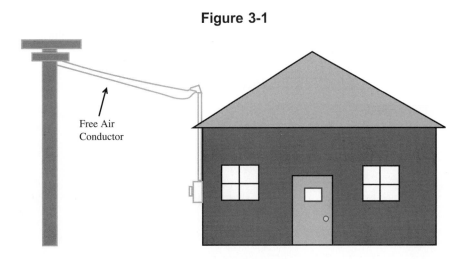

Free Air
Conductor

Terminal Ratings: 110.14(C)(1)

If a piece of equipment has terminals rated for 60°C, but 90°C conductors are used to supply the equipment, the ampacity of the conductors must not exceed 60°C. The lower rating is always used. Notice Figure 3-2. It shows terminals on a breaker that are rated for 75°C; therefore, the conductors connected to them are rated at 75°C.

Figure 3-2

Example 1

A welder has a terminal rated at 75°C and is supplied by #8 THHN conductors. What is the maximum ampacity of the conductors?

      50 Amps A #8 at 75°C can safely carry 50 amps

Ambient Temperature: A Practical Application

Ambient temperature is the temperature surrounding a particular conductor or piece of equipment. Improperly considering the ambient temperature around a conductor or equipment is one of the most common errors in an electrical application. For example, not many electricians consider the fact that when a conductor is run in an attic, the temperature during the summer months can exceed 120°F. Exposing a conductor to these high temperatures for an extended period of time can have a definite impact on the ampacity of the conductor. Excessive temperatures do not allow the conductor the necessary conditions for cooling and, therefore, the ampacity of the conductor is decreased. This can lead to circuit overloading. The information and example that follow will demonstrate the impact that ambient temperatures have on a conductor. Refer to Figure 3-3 for an example of conductors being installed in areas subject to high temperatures.

Figure 3-3

Derating or Correction Factors

Excessive heat can damage conductors, and proper measures should be used to factor in such occurrences. Correction factors are used to determine the proper ampacity under various temperature conditions. Table 310.16 (upper portion) considers only conductors used at 30°C or 86°F. To find the ampacity of a conductor other than 30°C or 86°F, use the bottom portion of Table 310.16 under the heading Correction Factors. Match the given temperature to the appropriate conductor column and multiply the ampacity by the correction factor. Table 3-2 shows the correction factors for the copper section only.

Table 3-2: Table 310.16 Correction Factors (Copper Conductors Only)

Ambient Temp. C	60C	75C	90C	Ambient Temp. F
21–25	1.08	1.05	1.04	70–77
26–30	1.00	1.00	1.00	78–86
31–35	0.91	0.94	0.96	87–95
36–40	0.82	0.88	0.91	96–104
41–45	0.71	0.82	0.87	105–113
46–50	0.58	0.75	0.82	114–122
51–55	0.41	0.67	0.76	123–131
56–60	-	0.58	0.71	132–140

(continued on next page)

Table 3-2 *(continued)*

Ambient Temp. C	60C	75C	90C	Ambient Temp. F
61–70	-	0.33	0.58	141–158
71–80	-	-	0.41	159–176

Example 1

What is the ampacity of a #3/0 THWN conductor installed in an attic with an ambient temperature of 150°F?

> First, use Table 310.16 to find the amperage for #3/0 THWN, which is 200 amps.
> Then, use the temperature correction factor of 0.33.

> $200 \times .33 = \textbf{66 A}$

> Remember, you must use the correction factor column where the wire type is located. For example, THWN is located in the 75°C column so the correction factor column must be 75°C.

Example 2

Eight #10 THHN conductors are run in a conduit at 31°C. What is the ampacity of each conductor?

> First, find the amperage of a #10 THHN, which is 30 amps because of 240.4(D).
> Then, find the correction factor for temperature, which is 0.96.
> Next, since there are more than three conductors in the conduit, use Table 310.15(B)(2)(a) for the adjustment factor, which is 70%.

> $30 \text{ A} \times 0.96 \times 70\% = \textbf{20.16 amps}$

Adjustment Factors: Table 310.15(B)(2)(a)

When more than three conductors are run in a raceway, heat again becomes a factor and must be calculated. As in the earlier cases, it is important to identify the current-carrying conductors. See Table 3-3.

Table 3-3: Table 310.15(B)(2)(a) Adjustment Factors for More Than Three Current-Carrying Conductors in a Raceway or Cable

Number of Current-Carrying Conductors	Percent of Values in Table 310.16 through 310.19 as Adjusted for Ambient Temperature if Necessary
4–6	80
7–9	70
10–20	50
21–30	45
31–40	40
41 and above	35

Example 1

A ¾" piece of EMT contains six #10 THHN conductors. What is the ampacity of the conductors under these conditions?

According to 240.4(D), a #10 can carry only 30 amps; 30 A × 80% = **24 amps**

Ampacity Practice Problems

1. What is the ampacity of a #12 THHW conductor installed in an area with an ambient temperature of 70°F?
   a. 15 A
   b. 20 A
   c. 26.25 A
   d. 21 A

2. Eight #6 TW conductors are installed in a conduit. What is the maximum ampacity of each conductor?
   a. 55 A
   b. 65 A
   c. 42 A
   d. 38.5 A

3. A conduit 24" long contains twelve #10 THHN conductors. What is the maximum ampacity of each conductor?
   a. 30 A
   b. 24 A
   c. 40 A
   d. 20 A

4. What is the allowable ampacity of a #6 THW conductor that is connected to a piece of equipment with terminals rated for 60°C?
   a. 55 A
   b. 65 A
   c. 50 A
   d. 60 A

5. A piece of EMT contains ten #6 THW conductors at a length of 30 feet. What is the allowable ampacity?
   a. 65 A
   b. 30.5 A
   c. 35 A
   d. 32.5 A

6. A conduit containing seven #12 THHN conductors is run through a store building attic with an ambient temperature of 110°F. What is the ampacity of the conductors?
   a. 17.4 A
   b. 12.18 A
   c. 20 A
   d. 30 A

7. What size overcurrent protection is required for six #14 THHN conductors in an ambient temperature of 104°F?
   a. 10.92 A
   b. 13.65 A
   c. 15 A
   d. 12.34 A

8. What is the ampacity of a #10 XHHW conductor installed in an ambient temperature of 86°F?
   a. 35 A
   b. 30 A
   c. 40 A
   d. 20 A

9. What is the ampacity of a 4/0 USE conductor run through an attic with an ambient temperature of 130°F?
   a. 230 A
   b. 154.1 A
   c. 164.5 A
   d. 210 A

10. A run of EMT contains eight #12 current-carrying conductors. What adjustment factor percentage should be applied to this situation?
    a. 80
    b. 70
    c. 50
    d. 45

Branch Circuit Ampacity and Loading, Article 210

This section of Chapter 3 discusses the application of certain branch circuit loads. Branch circuits are calculated differently depending upon what type of appliances or loads are connected to the circuit. The key words that follow help define some of different types of circuits.

Key Codes

Table 210.24

210.23(A) 80% rule and 50% rule, ex. AC window units

210.20(A) 125% rule of continuous load

How to Calculate the Minimum Number of Lighting Branch Circuits: Table 220.12

Determining the minimum number of lighting circuits is important in order to ensure that each lighting circuit is not overloaded. You should first determine the total lighting load of the structure based on the general lighting load by occupancy, located in Table 220.12. For example, a dwelling unit has a unit load of 3 volt amps per square foot, and a church would require 1 volt amp per square foot. Then calculate the lighting load by multiplying the unit load per square foot by the square footage of the structure based upon 220.12. Next calculate the circuit capacity by multiplying the circuit ampacity by 120 volts. For example, if lights are to be installed on a 15-amp circuit, you would multiply 15 × 120 volts to equal 1,800. If the lights are to be installed on a 20-amp circuit, you would multiply 20 × 120 volts to equal 2,400. Dividing the lighting load by the circuit capacity will then determine the minimum number of circuits required, but you will have to round up to the next whole number. The example that follows demonstrates this.

$$\text{Minimum number of branch circuits:} \quad \text{Number of circuits} = \frac{\text{lighting load}}{\text{circuit capacity}}$$

Example 1

How many 15-amp lighting circuits are required in a 2,000-square-foot home?

   Table 220.3(A) $3 \times 2,000 = \underline{6,000}$

   $15\ A \times 120\ V = 1,800\ \ 3.33$ or **4 circuits**

Notes on Branch Circuit and Load Terms

Some of the terminology used in determining branch circuit and loading calculations can be confusing. Terms such as **fixed appliance**, **portable appliance**, and **stationary appliance** are sometimes used to try to trick you on a question. You can find definitions of these terms in Article 550.2. Other terms that may be used are **general purpose branch circuits**, **individual branch circuits**, and **continuous load**. You should also be familiar with these terms, which are defined in Article 100.

Fixed appliance — An appliance that is secured in place at a certain location, such as a window A/C unit

Portable appliance — An appliance that can be easily moved or relocated, such as a blender. Refer to Figure 3-4 for more examples of portable appliances

Figure 3-4

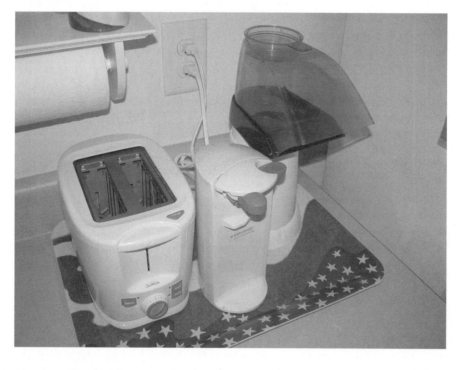

Stationary appliance — An appliance that cannot be easily moved or relocated, such as a refrigerator or range

General purpose branch circuits — Branch circuits used to provide power for general lighting and receptacle loads

Individual branch circuits — Branch circuits designated to supply one piece of equipment such as a welder, oven, or furnace

Continuous load — A load where the maximum current is expected to continue for three or more hours. *Receptacles are not considered to have continuous loads unless otherwise stated in the problem.*

Fastened in Place Equipment: 210.23(A)(2)

Fastened in place equipment, otherwise known as fixed appliances, are those pieces of equipment that are secured in a particular location. One of the most common examples is window air-conditioning units that are secured in the window casing or frame. Other examples of fastened or fixed appliances would include garbage disposals and trash compactors. Fixed appliances such as this cannot exceed 50% of the branch circuit rating according to 210.23(A)(2).

Example 1

Can a 12-amp A/C window unit that is fastened in place be installed on a 20-amp circuit?

> **No**. According to 210.23(A)(2), fastened in place equipment cannot exceed 50% of the branch circuit rating. Only 10 amps are allowed.

Continuous Duty: 210.19(A)(1)

Article 100 defines the term *continuous load* as one where the maximum current is expected to continue for three hours or more. Calculating continuous loads can be confusing. Article 210.19(A)(1) refers to branch circuit conductors and Article 210.20(A) refers to overcurrent protection. If you are trying to determine the demand on the overcurrent device for a continuous load, you must multiply the load by 125% and size the overcurrent device accordingly. In an attempt to simplify continuous load calculations, when the question asks for general lighting load or demand, you should multiply by 125%. Take a look at the examples to gain a better understanding of this calculation.

Example 1

What is the general lighting load for a 20-amp sign that is lit eight hours per day?

$$20 \text{ amps} \times 125\% = \textbf{25 amps}$$

This sign would require a 25 amp breaker instead of a 20 amp breaker because it is a continuous load. You may encounter problems that ask you to calculate circuit capacity in relation to continuous loads or to calculate the number of circuits required. In this case, you will multiply by 80%. Continuous loads can not exceed 80% of the branch circuit rating. Notice the question asks for circuit capacity and not general lighting load or demand.

Example 2

How much circuit capacity would a 120-volt, 15-amp breaker have if the lights connected on the circuit breaker operated for more than three consecutive hours?

$$15\text{-amp breaker} \times 120 \text{ volts} \times 80\% = \textbf{1,440 VA}$$

Notice in this equation that the total VA is reduced by multiplying by 80% in order to reduce the chance of the breaker overheating. In Example 2, you are being asked to calculate the capacity of the breaker based upon the current size breaker being used, the voltage, and the fact that the load in continuous. The capacity of the breaker is reduced from 1,800 VA to 1,440 VA as a result of the continuous load. Notice how the wording of the two examples changes the method by which the question is answered.

Branch Circuit and Conductor Ampacity Practice Problems

1. How many 15-amp lighting circuits are required in an 1,800-square-foot home?
   a. 1
   b. 2
   c. 3
   d. 4

2. There are 80 receptacles to be installed in an office building. How many 20-amp circuits should be installed? See 220.14 (K).
   a. 4
   b. 5
   c. 6
   d. 7

3. If a freezer is plugged into a 20-amp circuit and the load does not exceed 16 amps, can a small appliance be plugged into the same circuit as long as 20 amps is not exceeded?
   a. Yes
   b. No

4. What is the general lighting load for twenty-five 4' fluorescent fixtures, each containing one 1.8-amp ballast? Consider that the lights will operate at continuous duty.
   a. 45 A
   b. 50 A
   c. 52.75 A
   d. 56.25 A

5. Refer to question 4. How many 20-amps circuits are required for the lights?
   a. 1
   b. 2
   c. 3
   d. 4

6. How many receptacle outlets can be installed on a 20-amp branch circuit?
   a. 10
   b. 11
   c. 12
   d. 13

7. A disposal is added to an existing under-cabinet receptacle that is connected to three other receptacles. If the disposal is rated at 6 amps, can it be plugged into the receptacle, and if so, what is the maximum percentage of the circuit that the disposal can occupy?
   a. No, 50%
   b. Yes, 80%
   c. Yes, 50%
   d. No, 80%

8. Branch circuits that are 50 amps or larger are allowed to supply only _____ .
   a. lighting outlet loads
   b. nonlighting outlet loads
   c. motor loads
   d. loads of 40 amps or less

9. The maximum cord and plug connected load to a 20-amp breaker with a 15-amp rated receptacle is _____ .
   a. 12 amps
   b. 16 amps
   c. 20 amps
   d. 10 amps

10. The conductors that supply more than one receptacle for cord and plug portable loads shall have an ampacity of _____ the rating of the branch circuit.
    a. not less than 80%
    b. not greater than
    c. not less than
    d. none listed

Summary

Table 310.16 is one of the most used and also one of the most misused tables in the NEC. It is by far one of the most important tables due to the fact that it affects every conductor and branch circuit. This chapter also discussed the importance of derating and how ambient temperature plays a significant role in the ampacity of a conductor. It's imperative to have a good knowledge of how to calculate conductor and branch circuit ampacity, not only for the sake of the exam but even more importantly out in the field, because this addresses some major safety issues.

CHAPTER 4

CONDUIT AND BOX FILL CALCULATIONS

The following sections refer to properly calculating conduit and box fill. Oftentimes, these types of calculations are overlooked on the job site but they are very important. If a conduit or box is overfilled, excessive heat can build up and cause a reduction in the ampacity of the conductors. This can create some serious problems. Therefore, developing an understanding of how to calculate fill questions can lead to properly rated and loaded circuits.

Objectives

- Determine proper conduit fill
- Calculate proper box fill
- Identify electrical components for calculations

Conduit Fill Calculations

Conduit fill questions can be answered in two ways depending upon how the problem is set up. First, if the conductors in the conduit are all the same size, you can use Tables C.1 through C.12 in the NEC Annex to determine how many conductors are allowed in a certain tubing. You will need to know three things in order to use these tables: the type of conductor used, the type of tubing used, and the AWG size of the conductors. A second method may be necessary if the conductors are of different sizes. In this case, you will need to use Tables 4, 5, and 8 to determine the answer. The following details how to use both methods and the key code sections are listed below for conduit fill questions.

Key Codes

Chapter 9, Table 1
Chapter 9, Table 4
Chapter 9, Table 5
Chapter 9, Table 8
Annex C

Conductors of the Same Size

When the conductors in a certain type of conduit are all the same size, the calculation is relatively simple. Chapter 9, Tables C.1 through C.12 provide a quick solution to these types of problems.

Step 1. Determine how many conductors are in the raceway.

Step 2. Refer to Tables C1 through C12A in the Annex for the type of raceway or conduit used.

Step 3. Select size of the conductors.

Step 4. Select size of the raceway or conduit based upon the information in steps 1 through 3.

Example 1

Not more than five #10 THHN conductors are allowed in a ½" electrical metallic tubing (EMT).

Example 2

Not more than sixteen #12 THHN conductors are allowed in a ¾" EMT.

Example 3

Not more than twelve #8 THW conductors are allowed in a 1¼" intermediate metal conduit (IMC).

Table 4-1, which is an excerpt of NEC Table C.1, would be used when all the conductors are of the same size.

Table 4-1: Excerpt from Table C.1 Maximum Number of Conductors or Fixture Wires in Electrical Metallic Tubing (EMT) *(Based on Table 1, Chapter 9)*

		CONDUCTORS									
		Metric Designator (Trade Size)									
Type	**Conductor Size (AWG kcmil)**	**16** (1/2)	**21** (3/4)	**27** (1)	**35** (1 ¼)	**41** (1 ½)	**53** (2)	**63** (2 ½)	**78** (3)	**91** (3 ½)	**103** (4)
THHN,	14	12	22	35	61	84	138	241	364	476	608
THWN,	12	9	16	26	45	61	101	176	266	347	443
THWN-2	10	5	10	16	28	38	63	111	167	219	279
	8	3	6	9	16	22	36	64	96	126	161
	6	2	4	7	12	16	26	46	69	91	116

Reprinted with permission from the NFPA 70-2005, *National Electrical Code*®, Copyright © 2004, National Fire Protection Association, Quincy, MA 02269. This reprinted material is not the complete and official position of the NFPA on the referenced subject, which is represented only by the standard in its entirety.

Conductors of Different Sizes

If the conductors in the conduit are of different sizes, use an alternate method to calculate the fill. The following steps describe how to solve this type of problem.

Step 1. Refer to Chapter 9, Table 5.

Step 2. Find the size and type of conductors used to determine the cross-sectional area in square inches of the conductors.

Step 3. Refer to Chapter 9, Table 8 for area if bare conductors are used.

Step 4. Total the areas in square inches of all the conductors.

Step 5. Refer to Chapter 9, Table 4 for the type and size of the raceway.

Step 6. Find the minimum size allowed based on the total area and the allowable fill based upon the number of wires (1 wire, 2 wires, or over 2 wires).

Step 7. The sum must not be greater than the square inches allowed.

Example 1

Three #6 THW conductors and two #8 THW conductors are to be installed in EMT. What size should you use?

Chapter 9, Table 5   Three #6 THW $0.0726 \times 3 = .2178$

Two #8 THW $0.0437 \times 2 = \underline{.0874}$

0.3052

Select **1" EMT** from Table 4

Conduit Fill Practice Problems

1. Six #10 THWN conductors are to be installed in EMT. What size EMT should be used?
   a. ½"
   b. ¾"
   c. 1"
   d. 1½"

2. Four #2 TW conductors are to be installed in flexible metal conduit (FMC). What size should be used?
   a. ¾"
   b. 1"
   c. 1½"
   d. 1¼"

3. Ten #12 THHN conductors and four #10 THHN conductors are to be installed in (IMC). What size should you use?
   a. ½"
   b. ¾"
   c. 1"
   d. 1¼"

4. How many #10 THHN conductors can be installed in a 1½" rigid metal conduit (RMC) 24" long?
   a. 56
   b. 39
   c. 46
   d. 52

5. How many #14 TW conductors are allowed to be installed in ¾" schedule 80 rigid PVC conduit?
   a. 20
   b. 9
   c. 11
   d. 12

6. What is the percentage of cross-sectional area that can be used in a conduit when more than two conductors are installed?
   a. 53%
   b. 31%
   c. 40%
   d. none listed

7. How many #10 TW conductors can be installed in a section of ¾" liquid tight flexible nonmetallic conduit?
   a. 5
   b. 6
   c. 9
   d. 12

8. What is the total area (100%) of ½" EMT conduit?
   a. 0.304 sq in
   b. 0.433 sq in
   c. 0.622 sq in
   d. none listed

9. If a conduit nipple at a length of 24" or less is installed, it shall be filled to a maximum of _____ % of the cross-sectional area.
   a. 40
   b. 100
   c. 80
   d. 60

10. When conducting a calculation and the result contains a decimal, such as 5.3, 6.8, or 12.9, the next higher whole number shall be permitted to be used when the decimal number is _____ or greater.
    a. 0.5
    b. 0.8
    c. 0.6
    d. 0.7

Box Fill Calculations

As with conduit fill calculations, determining box fill depends upon whether the conductors are all the same size or different sizes. As with conduit fill, overcrowded boxes can lead to some major problems, so careful attention should be paid to box fill. The following are some important code references necessary to determine box fill.

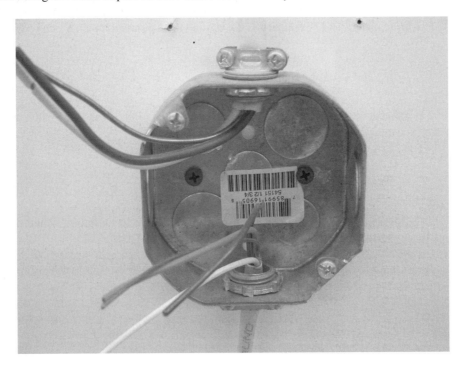

Key Codes

Table 314.16(A)
Table 314.16(B)
314.16(B)(1) through 314.16(B)(5)

Calculation Considerations

Before determining box fill, it is necessary to understand 314.16(B)(1) through 314.16(B)(5). This section of the NEC describes how each conductor, clamp, fitting, and device is calculated into the box fill equation. Consideration must be given to any object inside a box that takes up space and therefore reduces the available area for conductors. Be careful to read each question carefully and use only the information provided in the question. For example, if no clamps are mentioned in the question, do not assume any. Here is a brief description of how each is accounted for:

Conductors: Each conductor is counted once if it passes through, is terminated, or is spliced inside the box.

Clamps: Clamps should be counted once no matter how many are installed. In other words, if three clamps are installed in a box you would only count one. This should be based upon the largest conductor inside the box. See Figure 4-1.

Figure 4-1

Fittings: Hickeys and studs are counted once no matter how many are installed. This is based upon the largest conductor inside the box. See Figure 4-2.

Figure 4-2

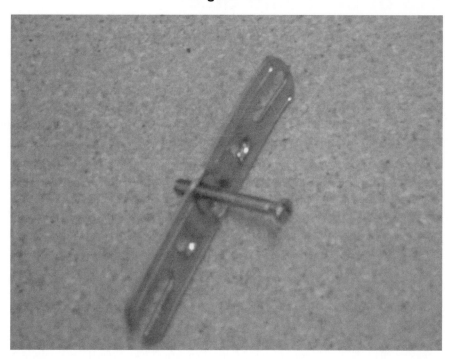

Devices: The yoke or strap of the device should be counted twice and based upon the largest conductor that is connected to the device. See Figure 4-3.

Figure 4-3

Equipment Grounding Conductor: Grounding conductors should be counted once no matter how many are installed in the box. This should be based upon the largest grounding conductor in the box.

Conductors of the Same Size

If the box contains conductors that are all the same size, you can use Table 314.16(A) to determine how many conductors are allowed for a specific box. However, before using the table, you must refer to Articles 314.16(B)(1) through 314.16(B)(5) to count the number of conductors. An excerpt from Table 314.16(A) is in Table 4-2, followed by an example that demonstrates how to use the information. Note that Table 314.16(A) is limited to metal boxes only. Nonmetallic boxes generally have the size stamped on the box itself for quick reference.

Table 4-2: Excerpt from Table 314.16(A) Metal Boxes

Box Trade Size In.		Minimum Volume		Maximum Number of Conductors*						
		Cm3	in.3	18	16	14	12	10	8	6
(3×2×1½")	device	123	7.5	5	4	3	3	3	2	1
(3×2×2)	device	164	10.0	6	5	5	4	4	3	2
(3×2×2¼")	device	172	10.5	7	6	5	4	4	3	2
(3×2×2½")	device	205	12.5	8	7	6	5	5	4	2
(3×2×2¾")	device	230	14.0	9	8	7	6	5	4	2
(3×2×3½")	device	295	18.0	12	10	9	8	7	6	3

Example 1

A 3"× 2"× 2½" device box holds the following: six #12 conductors, three grounding conductors, four cable clamps, and one duplex receptacle. Does the box meet Code?

Six #12 conductors	6
Three grounds	1 **No, Code allows only 5**
Four clamps	1
One receptacle	2
	10 Total

Example 2

What size square box would allow for ten #8 conductors, three grounds, and one receptacle?

Ten #8 conductors	10
Three grounds	1
One receptacle	2
	13 Total

(4 11/16" × 2 1/8") square
None of the other square boxes will accommodate thirteen #8 conductors.

Conductors of Different Sizes

When you calculate box fill and use different sizes of conductors, you must use an alternate method to find the solution. An additional table, Table 314.16(B), is needed in conjunction with Table 314.16(A). When different sizes of conductors share a box, it is required that the volume of each conductor size be factored into the equation. Table 314.16(B), found in Table 4-3, provides the information necessary to complete such calculations. Some example problems follow.

Table 4-3: Table 314.16(B) Volume Allowance Required per Conductor

	Free Space Within Box for Each Conductor	
Size of Conductor (AWG)	cm3	in3
18	24.6	1.50
16	28.7	1.75
14	32.8	2.00
12	36.9	2.25
10	41.0	2.50
8	49.2	3.00
6	81.9	5.00

Reprinted with permission from the NFPA 70-2005, *National Electrical Code*®, Copyright © 2004, National Fire Protection Association, Quincy, MA 02269. This reprinted material is not the complete and official position of the NFPA on the referenced subject, which is represented only by the standard in its entirety.

Example 1

What is the minimum cubic-inch volume required for a box containing two clamps, one switch, and two #14 and two #12 conductors with grounds?

Two #14 conductors	$2.00 \times 2 = 4.00$
Two #12 conductors	$2.25 \times 2 = 4.50$
Two clamps	$2.25 \times 1 = 2.25$
One switch	$2.25 \times 2 = 4.50$
Two grounds	$2.25 \times 1 = \underline{2.25}$

17.50 cubic inches

Next, find a device box that will accommodate 17.50 cubic inches, such as a ($3" \times 2" \times 3\ 1/2"$) device box which has a volume of 18.0 cubic inches.

Example 2

A certain box contains a #12/2 sheathed cable and a #14/2 sheathed cable. The #12 conductors are connected to a duplex receptacle and the #14 conductors to a single pole switch. The conductors are held in place by two cable clamps. What is the minimum cubic-inch volume requirement?

Two #12 conductors	$2.25 \times 2 = 4.50$
Two #14 conductors	$2.00 \times 2 = 4.00$
Two grounds	$2.25 \times 1 = 2.25$
Two clamps	$2.25 \times 1 = 2.25$
One switch	$2.25 \times 2 = 4.50$
Two grounds	$2.00 \times 2 = \underline{4.00}$

21.5 cubic inches

Note that each sheathed cable contains one black, one white, and one grounding conductor, so be sure to read the question carefully. The receptacle and switch are counted based upon the size of conductor connected to each.

Box Fill Practice Problems

1. How many #14 conductors are allowed to be installed in a ($3" \times 2" \times 2"$) device box?
   a. 3
   b. 4
   c. 5
   d. 6

2. What is the minimum cubic volume allowance for a ($4" \times 2\frac{1}{8}"$) round box?
   a. 15.5
   b. 18.0
   c. 12.5
   d. 21.5

3. What size device box should be used for six #12 conductors? The box also contains two grounding conductors.
   a. ($3" \times 2" \times 3\frac{1}{2}"$)
   b. ($3" \times 2" \times 2"$)
   c. ($3" \times 2" \times 2\frac{1}{2}"$)
   d. ($3" \times 2" \times 2\frac{1}{4}"$)

4. How many cubic inches are required for four #14 conductors, two #12 conductors, two grounding conductors, and four clamps?
   a. 15.5
   b. 16.25
   c. 14.0
   d. 17.0

5. How many cubic inches are required for two #14 conductors connected to a switch, two #12 conductors connected to a receptacle, two grounding conductors, two clamps, and one stud?
   a. 17.25
   b. 19.5
   c. 21.0
   d. 23.75

6. Each conductor that passes through a box without a splice or termination shall be counted _____ time(s).
   a. 1
   b. 2
   c. 1½
   d. 0

7. A single pole receptacle installed in a box shall be calculated based upon _____ .
   a. single volume based on the largest conductor connected to it
   b. double volume based on the largest conductor in the box
   c. double volume based on the largest conductor connected to it
   d. single volume based on the largest conductor in the box

Summary

This chapter has covered some of the most common calculations performed in the field and also some of the most overlooked calculations, which make for great exam questions. Most exams ask conduit and box fill calculation questions because it is easy to get lost in all the tables as well as all of the different steps involved in obtaining a correct answer. Therefore, take you time and make sure that you do not get in too big a hurry when using tables, which play a key role in determining conduit and box fill.

CHAPTER 5

RESIDENTIAL LOAD CALCULATIONS

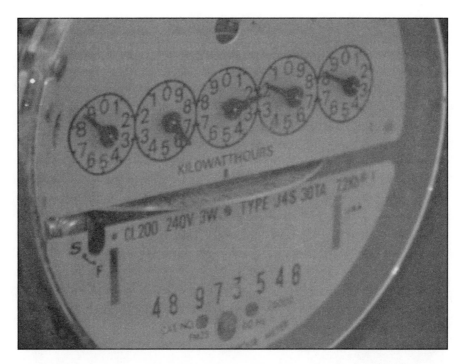

Since most exams will probably not ask you to perform complete load calculations, this chapter will cover several components of the load calculation. For example, you may be asked a question concerning just dryers or ranges. This chapter will take a look at several tables and sections 220.52, 220.53, 220.54, and 220.54, and how to use each to determine a particular load. Load calculations can be some of the more difficult calculations to perform because many errors can be made, so take your time when answering these questions. Also, remember that some sample load calculations are provided in the back of the NEC for reference.

Objectives

- Calculate lighting demand for a residence
- Determine demand for residential cooking equipment
- Calculate appliance load demand and use other helpful tables

Lighting Demand: Article 220.12 and Table 220.42

Calculating lighting demand involves several steps to complete. You first need to determine how to calculate the total square footage of the home. Article 220.12 defines what portions of the residence are to be included in the calculations and what portions are to be excluded. Floor plans like the ones shown in Figure 5-1 are generally used to calculate square footage.

Figure 5-1

Table 220-12 provides the unit load for particular occupancies. For a residential occupancy (dwelling unit), the unit load is 3 volt amps per square foot. The following example demonstrates this 3 VA multiplied by the square footage.

$$1,400 \text{ sq ft residence} \times 3 \text{ VA} = \mathbf{4,200\ VA}$$

However, when calculating the general lighting load you must not only consider the square footage but you must add the two required small-appliance circuits (210.11 (C)(1)) and the one required laundry circuit (210.11(C)(3)). Each of these circuits must be assigned 1,500 volt amps according to 220.52(A) and 220.52(B). Now the general lighting load for the residence above can be calculated.

$$1,400 \text{ sq ft residence} \times 3 \text{ VA} = 4,200 \text{ VA}$$
$$2 \text{ small-appliance circuits} \times 1,500 \text{ VA} = 3,000 \text{ VA}$$
$$1 \text{ laundry circuit} \times 1,500 \text{ VA} = \underline{1,500 \text{ VA}}$$
$$\text{Total} \quad \mathbf{8,700\ VA}$$

Finally, the NEC allows for demand factors to be applied to the calculation above. So far, the calculation assumes that every light as well as all the small-appliance and laundry circuits are being utilized all the time. Under normal operating conditions, all the lights will not be on and every receptacle will not be used 100% of the time. Therefore, the NEC allows the calculation above to be reduced according to Table 220.42. Refer to Table 5-1 to determine how to apply the demand factors.

$$1,400 \text{ sq ft residence} \times 3 \text{ VA} = 4,200 \text{ VA}$$
$$2 \text{ small-appliance circuits} \times 1,500 \text{ VA} = 3,000 \text{ VA}$$
$$1 \text{ laundry circuit} \times 1,500 \text{ VA} = \underline{1,500 \text{ VA}}$$

Total   **8,700 VA**

$$\text{First } 3,000 @ 100\% =$$
$$(3,000 \times 100\%) = 3,000 \text{ VA}$$
$$\text{From } 3,001 \text{ to } 120,000 @ 35\% =$$
$$(8,700 - 3,000 = 5,700 \times 35\% = 1,995) = \underline{1,995 \text{ VA}}$$

Total   **4,995 VA**

Table 5-1: Excerpt from Table 220.42 Lighting Load Demand Factors

Type of Occupancy	Portion of Lighting Load to Which Demand Factor Applies (Volt-Amperes)	Demand Factor (Percent)
Dwelling units	First 3000 or less at	100
	From 3001 to120,000 at	35
	Remainder over 120,000 at	25

Reprinted with permission from the NFPA 70-2005, *National Electrical Code*®, Copyright © 2004, National Fire Protection Association, Quincy, MA 02269. This reprinted material is not the complete and official position of the NFPA on the referenced subject, which is represented only by the standard in its entirety.

Lighting Demand Practice Problems

1. Calculate the total lighting demand for a 1,700-square-foot home. The home should contain the minimum two small-appliance circuits and one laundry circuit.
   a. 9,600 VA
   b. 8,410 VA
   c. 5,310 VA
   d. none listed

2. A certain residence has 1,400 square feet of living space, a 400-square-foot porch, and a 500-square-foot garage. Calculate the lighting demand. The home should contain the minimum two small-appliance circuits and one laundry circuit.
   a. 4,995 VA
   b. 8,700 VA
   c. 5,940 VA
   d. none listed

3. A house contains 1,850 square feet in which 600 square feet is the unfinished basement. Calculate the total lighting demand for the house. The home should contain the minimum two small-appliance circuits and one laundry circuit.
   a. 5,468 VA
   b. 4,838 VA
   c. 8,250 VA
   d. none listed

Cooktops, Ovens, and Ranges: Article 220.55 and Table 220.55

Cooktops, ovens, and ranges are calculated using Table 220.55. The following examples demonstrate how to calculate each. Notice that Column C can be used in lieu of Column A or B. This will be demonstrated later. Refer to Table 5-2.

Table 5-2: Table 220.55 Demand Factors and Loads

Number of Appliances	Demand Factor % Column A (Less than 3 ½ kW Rating)	Demand Factor % Column B (3 ½ kW to 8 ¾ kW Rating)	Column C Maximum Demand (kW) (See Notes) (Not over 12 kW Rating)
1	80	80	8
2	75	65	11
3	70	55	14
4	66	50	17
5	62	45	20
6	59	43	21
7	56	40	22
8	53	36	23
9	51	35	24
10	49	34	25
11	47	32	26
12	45	32	27
13	43	32	28
14	41	32	29
15	40	32	30
16	39	28	31
17	38	28	32
18	37	28	33
19	36	28	24
20	35	28	35
21	34	26	36
22	33	26	37
23	32	26	38
24	31	26	39
25	30	26	40

(continued on next page)

Table 5-2 *(continued)*

| | Demand Factor % | Demand Factor % | |
Number of Appliances	Column A (Less than 3 ½ kW Rating)	Column B (3 ½ kW to 8 ¾ kW Rating)	Column C Maximum Demand (kW) (See Notes) (Not over 12 kW Rating)
26–30	30	24	15 kW + 1 kW for each range
31–40	30	22	
41–50	30	20	25 kW + ¾ kW for each range
51–60	30	18	
61 and over	30	16	

Reprinted with permission from the NFPA 70-2005, *National Electrical Code*®, Copyright © 2004, National Fire Protection Association, Quincy, MA 02269. This reprinted material is not the complete and official position of the NFPA on the referenced subject, which is represented only by the standard in its entirety.

The following examples are used to determine the service demand load from Table 220.55. Pay attention to Note 1 at the bottom of Table 220.55 for instructions on the use of Column C. A later discussion on the use of Note 4 will concern how to determine the branch circuit load.

Single Column Usage Examples

Example 1

What is the service demand for a 3 kW cooktop?

> 80 % of 3 kW or **2.4KW**   Use Col. A

Example 2

What is the service demand for a 6 kW cooktop?

> 80 % of 6 kW or **4.8 kW**   Use Col. B

Example 3

What is the service demand for two 10 kW ovens?

> **11 kW**   Use Col. C

Example 4

What is the service demand for a 14 kW range?

> 14 kW − 12 kW = 2 kW over

> Since 14 kW is larger than the maximum of 12 kW per Column C, you must add 5% for each additional kW over 12. See Note 1 under Table 220.55.

> First take the 8 kW given in the table for one range under Column C, then add 10% to 8 kW to get the total.

> 8 kW + 10% = **8.8 kW**

Multiple Column Usage Examples

When a calculation involves values from two different columns, you need to find the percentage in the appropriate column for each unit, then add the values together. There may be some cases where you will need to choose between uses from two different columns. Most exams ask for the *lowest* demand allowed. In that case, you will need to compare the columns to determine the lowest allowed demand. The examples below demonstrate how. Column C can be used as an alternative to Columns A and B.

Example 1

What is the service demand for one 3½ kW cooktop and one 6 kW oven?

$$3\tfrac{1}{2} \text{ kW @ } 80\% = 2.8 \text{ kW} \qquad \text{Use Col. A}$$
$$6 \text{ kW @ } 80\% = \underline{4.8 \text{ kW}} \qquad \text{Use Col. B}$$
$$\text{Total} \quad \textbf{7.6 kW}$$

Using Column C would produce an answer of 11 kW, which is more than 7.6 kW. Therefore, stick with your first answer of 7.6 kW, since it is the lowest.

Example 2

What is the demand for an 11 kW oven and a 5 kW range?

First, calculate each under the appropriate column:

$$11 \text{ kW @ } \qquad 8 \text{ kW} \qquad \text{Use Col. C}$$
$$5 \text{ kW @ } 80\% = \underline{4 \text{ kW}} \qquad \text{Use Col. B}$$
$$\text{Total} \quad \textbf{12 kW}$$

However, check to see if Column C produces a smaller demand:
Two units, using Column C, would be 11 kW. Therefore, use the Column C demand of 11 kW since it is less than 12 kW.
The correct answer is 11 kW using Col. C.

Branch Circuit Load

Some exams may specify that you should calculate the branch circuit demand and not the service or feeder demand. If the question states that the branch circuit demand should be used, you should apply Note 4 for Table 220.55. In this case, the NEC allows you to add all of the appliances together and treat them as one appliance. The example below demonstrates.

Example 1

What is the branch circuit demand for two 6 kW ovens and one 4 kW cooktop?

Add 6 kW + 6 kW + 4 kW = 16 kW   then, using Column C:
8 kW plus 5% for each additional kW over 12:
8 kW + 20% = **9.6 kW**

Cooktops, Ovens, and Ranges Practice Problems

1. What is the service demand for three 3 ½ kW cooktops?
   a. 3.5 kW
   b. 7.35 kW
   c. 8 kW
   d. none listed

2. What is the service demand for one 13 kW oven?
   a. 8.4 kW
   b. 8 kW
   c. 11 kW
   d. none listed

3. What is the service demand for six 8 kW ranges?
   a. 40 kW
   b. 21.5 kW
   c. 17.2 kW
   d. none listed

4. What is the service demand for one 3½ kW cooktop and one 7 kW oven?
   a. 10.5 kW
   b. 8.4 kW
   c. 11 kW
   d. none listed

5. What is the branch circuit demand for two 4 kW cooktops and one 6 kW cooktop?
   a. 8 kW
   b. 8.8 kW
   c. 11 kW
   d. none listed

6. What is the service demand for two 3 kW ranges and two 5 kW ovens?
   a. 8 kW
   b. 16 kW
   c. 8.8 kW
   d. none listed

Dryers: Article 220.54 and Table 220.54

Figure 5-2 illustrates a common dryer receptacle. Calculating dryers for residential installations is fairly easy. However, you should keep in mind Article 220.54, which states that you should use either 5,000 watts or the nameplate rating of the dryer, whichever is larger. Refer to Table 5-3.

Figure 5-2

Table 5-3: Table 220.54 Demand Factors for Household Electric Clothes Dryers

Number of Dryers	Demand Factor (Percent)
1–4	100%
5	85%
6	75%
7	65%
8	60%
9	55%
10	50%
11	47%
12–22	% = 47 – (number of dryers – 11)
23	35%
24–42	% = 35 – [0.5 × (number of dryers - 23)]
43 and over	25%

Example 1

What is the demand for a 6 kW dryer?

Using Table 220.54, the answer would be **6 kW**.

6 kW @ 100% = **6 kW**

Example 2

What is the demand for seven 5,000 watt dryers?

7×5 kW = 35 kW @ 65% = **22.75 kW**

Example 3

What is the demand for a 4 kW dryer?

Since the 5,000 watt demand is the smallest demand according to Article 220.54, **5 kW** should be used.

Dryers Practice Problems

1. What is the demand for five 6 kW dryers?
   a. 30,000 W
   b. 25,500 W
   c. 22,500 W
   d. none listed

2. What is the demand for four 3.5 kW dryers?
   a. 14,000 W
   b. 17,000 W
   c. 20,000 W
   d. none listed

3. What is the demand for ten 6,500-watt clothes dryers?
   a. 32,500 W
   b. 65,000 W
   c. 35,750 W
   d. none listed

4. Twenty 5 kW clothes dryers are to be installed in an apartment building. What is the demand?
   a. 100 kW
   b. 57 kW
   c. 32 kW
   d. none listed

5. A total of forty-four 4.5 kW dryers are installed in a college dorm. What is the demand?
   a. 49.5 kW
   b. 55 kW
   c. 65 kW
   d. none listed

Appliance Load: Article 220.53

Additional questions may be asked concerning calculating the appliance load of a residence. According to 220.53, the NEC allows for a demand factor of 75% to be applied if the number of appliances is four or more. However, you should be aware of the appliances that do not apply to the 75% demand factor such as dryers, ranges, space heaters, and air conditioners.

Example 1

The following appliances are to be installed in a residence:

6 kW water heater

120 volt, 3 amp trash compactor

120 volt, 2 amp disposal

5 kW dishwasher

What is the demand?

Water heater $\quad= 6,000$ watts

Compactor $= 120 \times 3 = \quad 360$ watts

Disposal $= 120 \times 2 \quad = \quad 240$ watts

Dishwasher $\quad = \underline{5,000}$ watts

$11,600$ watts @ $75\% = \textbf{8,700 watts}$

Apply the 75% demand factor, since there are four appliances.

Appliance Load Practice Problems

Find the demand for the following list of appliances.

1. 4 kW dishwasher
   240 volt, 4 amp pool pump
   5 kW water heater
   6 kW dryer

   a. 11,970 watts
   b. 9,960 watts
   c. 7,470 watts
   d. 10,574 watts

2. 120 volt, 2.5 amp disposal
   5 kW water heater
   120 volt, 3 amp trash compactor
   240 volt, 5 amp pool pump
   4 kW dishwasher

   a. 8,145 watts
   b. 10,860 watts
   c. 9,185 watts
   d. 7,245 watts

Grounding Electrode Conductor Sizing

There are some additional tables that may be referred to on an exam, such as Table 5-4. The questions will probably ask you to use given information and a certain table to find the correct answer. The following are some examples.

Table 250.66 would be used to determine the proper size grounding electrode conductor for A/C systems. You may encounter a question that provides the service entrance conductor size and wants you to determine the correct size grounding electrode conductor size. Figure 5-3 demonstrates an underground service in which a grounding electrode conductor would be sized.

Figure 5-3

Table 5-4: Table 250.66 Grounding Electrode Conductor for Alternating – Current Systems

| Size of Largest Ungrounded Service-Entrance Conductor or Equivalent Area for Parallel Conductors (AWG/kcmil) | | Size of Grounding Electrode Conductor (AWG/kcmil) | |
Copper	Aluminum or Copper-Clad Aluminum	Copper	Aluminum or Copper-Clad Aluminum
2 or smaller	1/0 or smaller	8	6
1 or 1/0	2/0 or 3/0	6	4
2/0 or 3/0	4/0 or 250	4	2
Over 3/0 through 350	Over 250 through 500	2	1/0
Over 350 through 600	Over 500 through 900	1/0	3/0
Over 600 through 1100	Over 900 through 1750	2/0	4/0
Over 1100	Over 1750	3/0	250

Reprinted with permission from the NFPA 70-2005, *National Electrical Code*®, Copyright © 2004, National Fire Protection Association, Quincy, MA 02269. This reprinted material is not the complete and official position of the NFPA on the referenced subject, which is represented only by the standard in its entirety.

Example 1

What size copper grounding electrode conductor is needed for a service fitted with a copper 3/0 service entrance cable?

> First, locate the service entrance conductor size under the first copper column. Then, follow that row across to the grounding electrode copper column. **The answer is a #4.**

Example 2

What size aluminum grounding electrode conductor is needed for a service built with a 4/0 aluminum service conductor? **The answer is a #2.**

Grounding Electrode Conductor Practice Problems

1. What size copper grounding electrode conductor is required for a #1 copper service entrance conductor?
   a. 8
   b. 6
   c. 4
   d. 2

2. If no service conductors are present, the grounding electrode conductor should be sized according to _____ .
   a. the combined nameplate ratings of all motor loads
   b. the equivalent size of the largest service entrance conductor required for the served load
   c. the largest conductor utilized in the facility
   d. none of the above

3. What size aluminum grounding electrode conductor is required for a 2/0 aluminum service entrance conductor?
   a. 6
   b. 4
   c. 2
   d. 1/0

Summary

Residential load calculations make for some great exam questions. There are many different parts to residential calculations and the subject encompasses a tremendous amount of material. These types of calculations require the test taker to be familiar with Article 220 and also well rounded in basic math skills. This chapter has provided you with a good overview of the many different types of residential questions that you may encounter. Pay close attention to key words contained in each question, because little subtle words can make a big difference in the correct answer. Again, table usage plays an important role in determining many of the solutions to residential load calculations, so approach each table with caution and read them carefully.

CHAPTER 6

COMMERCIAL LOAD CALCULATIONS

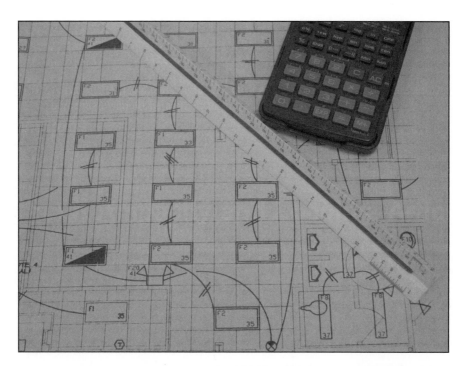

When calculating commercial load demands, it is important to understand the concept of a continuous load, which is when the load is expected to operate for more than three hours. If a load is considered continuous, then the demand of the load must be increased by 25% according to Article 215.2(A)(1) and Article 230.42 (A)(1). The following examples demonstrate certain portions of commercial load calculations. This chapter includes calculating lighting demand, receptacles, show window lighting, and commercial cooking equipment.

Objectives

- Determine the general lighting demand for commercial applications

- Calculate receptacle, multioutlet assemblies, and show window demand

- Use tables for determining commercial cooking equipment demand and sizing the equipment grounding conductor.

General Lighting Demand: Table 220.12 and Table 220.42

In order to properly calculate lighting demand, you must first determine the total VA using the square footage provided and multiply it by the appropriate VA in Table 220.12. Refer to Table 6-1. Then, use Table 220.42 to determine if a demand factor applies to the type of structure involved in the calculation. The following examples show how to use this information.

Table 6-1: Excerpt from Table 220.12 General Lighting Loads by Occupancy

Type of Occupancy	Unit Load	
	Volt-Amperes Per Square Meter	Volt-Amperes per Square Meter
Armories and auditoriums	11	1
Banks	39[b]	3½[b]
Barber shops and beauty parlors	33	3
Churches	11	1
Clubs	22	2
Courtrooms	22	2

[b] See 220.14(K)

Reprinted with permission from the NFPA 70-2005, *National Electrical Code*®, Copyright © 2004, National Fire Protection Association, Quincy, MA 02269. This reprinted material is not the complete and official position of the NFPA on the referenced subject, which is represented only by the standard in its entirety.

Example 1

What is the lighting demand for a 7,000-square-foot office building?

Table 220.12 $7,000 \times 3.5 \text{ VA} \times 125\% = \textbf{30,625 VA}$

The 125% must be used due to the fact that office lighting will certainly be used for more than three hours at a time.
Since office buildings do not have a specific category in Table 220.42, you should use the "All others" category, which says to take 100% of the VA.

Example 2

What is the general lighting demand for a 5,000-square-foot warehouse?

Table 220.12 $5,000 \times \frac{1}{4} \text{ VA} \times 125\% = \textbf{1,563 VA}$

The 125% must be used due to the fact that warehouse lighting will certainly be used for more than three hours at a time.
According to Table 220.42, since the 1,563 VA does not exceed 12,500 VA, you should take the value at 100%.

Example 3

What is the lighting demand for a motel with 6,000 square feet of rooms and 2,000 square feet of hallway?

Table 220.12	$6,000 \times 2$ VA	$= 12,000$ VA for rooms	
	$2,000 \times \frac{1}{2}$ VA $\times 125\%$	$= 1,250$ VA for the hallway	

Rooms 6,000 VA@50% =	3,000 VA per Table 220.42
Hallway	+1,250 VA
	4,250 VA

Since most motel room lighting is not considered a continuous load, the 125% was not applied to that portion of the equation. However, hallway lights are continuously on and the 125% must be applied.

Commercial Lighting Load Practice Calculations

1. What is the lighting demand for a 10,000-square-foot office complex?
   a. 35,000 VA
   b. 43,750 VA
   c. 30,000 VA
   d. none listed

2. What is the lighting demand for a 5,500-square-foot bank?
   a. 24,063 VA
   b. 19,250 VA
   c. 22,560 VA
   d. none listed

3. What is the lighting demand for a 12,000-square-foot church?
   a. 12,000 VA
   b. 14,000 VA
   c. 21,000 VA
   d. none listed

4. What is the lighting demand for a hotel with 10,000 square feet of guest rooms and 2,000 square feet of hallway?
   a. 21,125 VA
   b. 21,000 VA
   c. 10,450 VA
   d. 10,625 VA

5. A hotel is being built with 125 rooms, each measuring 15' × 25'. The hotel will also have seven hallways measuring 75' × 10'. What is the lighting demand?
   a. 42,125 VA
   b. 100,313 VA
   c. 50,157 VA
   d. 90,180 VA

Receptacles, Multioutlet Assemblies, and Show Window Lighting: Article 220.14

Some good test questions that may be asked on the exam refer to commercial building receptacles, mulitoutlet assemblies, and show window lighting. These calculations will require referring to sections of Article 220.14 and Table 220.44. The examples below demonstrate each of these scenarios.

Receptacles

The NEC allows demand factors to be applied when calculating nondwelling receptacle loads. The first 10 kVA is taken at 100% and the remainder over 10 kVA is taken at 50%. When receptacle problems are asked, they may appear in two major forms. The questions could contain information given in kVA. If that is the case, you should refer to Table 220.44. The question could also be stated using the number of receptacles installed. In that case, you should refer to Article 220.14 (L). This sections states that each outlet should be based upon 180 volt amps each. Then, you would need to refer to Table 220.44 to apply the demand factors.

Example 1

An office complex contains 40 kVA of receptacles. What is the demand?

> Table 220.44   First 10 kVA @ 100%   = 10 kVA
> Remainder (30 kVA) @ 50% = 15 kVA
> = **25 kVA**

Example 2

A bank has 120 receptacles that are to be installed. What is the demand?

> 220.14 (L)   180 volt amps per receptacle
> $180 \times 120 = 21,600$ VA or 21.6 kVA

> Table 220.44   First 10 kVA @ 100%   = 10 kVA
> Remainder (11.6 kVA) @ 50% = 5.8 kVA
> = **15.8 kVA**

Multioutlet Assemblies

According to Article 220.14 (H), each 5 feet of multioutlet assemblies is to be calculated at 180 VA. Multioutlet assemblies are routinely used in shop areas or other places that require a number of receptacles in a certain given space. See Figure 6-1.

Example 1

A continuous length of a multioutlet assembly measuring 25 feet is installed in a shop. What are the calculated volt amps?

> 220.14 (H) $25 / 5 = 5 \times 180$ VA = **900 VA**

Example 2

Referring to the previous example, what would be the demand in amps?

> 900VA / 120V = **7.5 amps**

Figure 6-1

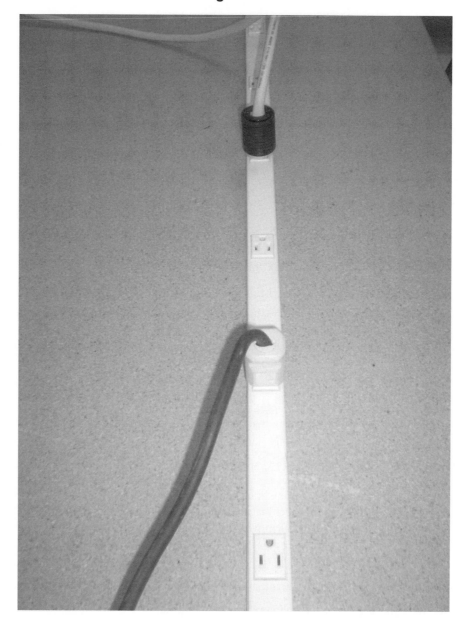

Show Window Lighting

Article 220.14 (G) of the NEC states that show window lighting shall be calculated at 200 VA for every 1 foot of lighting. You should also consider whether or not the show window lighting will be used for more than three continuous hours. If so, you must add 125% to the calculation. In order to determine the load, simply multiply 200 volt amps times the linear footage. See Figure 6-2.

Figure 6-2

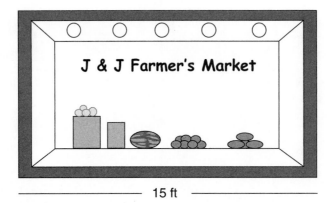

Example 1

Using Figure 6-2, what is the lighting load demand for this example?

$$200 \text{ volt amps} \times 15 \text{ ft} = \textbf{3,000 VA}$$

Notice that the question did not mention how many hours the show window lighting would be operating, so you should assume that it is less than three hours of continous lighting. If the question had stated the number of hours the show window lighting would be used and if it had been more than three hours, you would have had to factor in 125%. The next example demonstrates a continuous load.

Example 2

A clothing store has 35 feet of show window lighting that is operated a minimum of eight hours per day. What is the demand?

$$220.14 \text{ (G)} \quad 35' \times 200 \text{ VA} = 7,000 \text{ VA} \times 125\% = \textbf{8,750 VA}$$

Some questions may ask you to calculate how many branch circuits are required for a particular section of show window lighting. If a question like this should arise, remember that a circuit can be loaded to only 80% of the rating. Take a look at the following example to see how this type of problem would be solved.

Example 3

How many 20-amp circuits would be needed for 60' of show window lighting?

$$220.14 \text{ (G)} \quad \frac{60' \times 200 \text{ VA}}{20 \text{ A} \times 120 \text{ V} \times 80\% = 1,920 \text{VA}} = \frac{12,000 \text{ VA}}{} = \textbf{6.25 or 7 circuits}$$

Receptacles, Multioutlet Assemblies, and Show Window Lighting Practice Problems

1. An insurance office is being constructed and will have 65 receptacles installed. What is the demand?
   a. 11.7 kVA
   b. 10.85 kVA
   c. 12.5 kVA
   d. 10 kVA

2. A bank building has 145 receptacles installed. What is the demand?
   a. (A) 26.1 kVA
   b. (B) 24 kVA
   c. (C) 20.5 kVA
   d. (D) 18.05 kVA

3. A 12' continuous length of multioutlet assembly is installed in an electronics shop. What is the demand?
   a. 2.4 amps
   b. 3.6 amps
   c. 6 amps
   d. 12 amps

4. Three continuous lengths of multioutlet strips each measuring 10' are installed. What is the total demand in amps?
   a. 9 amps
   b. 3 amps
   c. 12 amps
   d. 10 amps

5. A shoe store has 30' of show window lighting. The store plans to use the lighting for 10 hours per day. What is the demand?
   a. 6,000 VA
   b. 4,500 VA
   c. 7,500 VA
   d. 5,000 VA

Commercial Cooking Equipment: Article 220.56 and Table 220.56

Article 220.56 allows for demand factors to be applied to commercial cooking equipment and related equipment such as water heaters and booster heaters when three or more are used. However, the demand factor calculation should not be lower than the sum of the two largest pieces of equipment. The demand factor information can be found in Table 220.56, which is shown in Table 6-2.

Table 6-2: Table 220.56: Demand Factors for Kitchen Equipment – Other Than Dwelling Unit(s)

Number of Units of Equipment	Demand Factor (Percent)
1	100
2	100
3	90
4	80
5	70
6 and over	65

Example 1

The following kitchen equipment is installed in a school. What is the demand?

> 1 – 8,000 VA oven
> 1 – 7,000 VA dishwasher
> 1 – 7,500 VA stove
> 1 – 6,000 VA water heater

$$8,000 + 7,000 + 7,500 + 6,000 = 28,500 \times 80\% = \mathbf{22,800\ VA}$$

Commercial Cooking Equipment Practice Problems

1. A small restaurant has a 6 kVA oven, a 5 kVA water heater, and a 5.5 kVA stove. What is the demand?
   a. 14.85 kVA
   b. 16 kVA
   c. 18.5 kVA
   d. 14 kVA

2. A school cafeteria has a 5.5 kVA dishwasher, two 6 kVA ovens, two 4.5 kVA stoves, a 4 kVA deep fryer, and a 5 kVA water heater. What is the demand?
   a. 35.5 kVA
   b. 21.4 kVA
   c. 23.08 kVA
   d. 27.3 kVA

3. What is the demand factor percentage to be used when a commercial kitchen contains one 220 V coffee maker, one 8 kVA oven, one 6.5 kVA dishwasher, and one gas stove?
   a. 100%
   b. 90%
   c. 80%
   d. 70%

4. What is the demand for one 10 kW oven and one 8 kW range?
   a. 16.2 kW
   b. 18 kW
   c. 15 kW
   d. 20 kW

5. According to Article 220.56, the feeder demand shall not be _____ the sum of the two _____ kitchen equipment loads.
   a. less than, largest
   b. less than, smallest
   c. greater than, largest
   d. greater than, smallest

6. What is the feeder demand for a commercial kitchen that contains the following:
   1 – 10 kW stove
   1 – 5 kW coffee maker
   1 – 8 kW dishwasher
   1 – 7.5 kW french fry cooker
   a. 30.5 kW
   b. 24.4 kW
   c. 40 kW
   d. 27.45 kW

Sizing the Equipment Grounding Conductor

As in the previous chapter, there may be questions on the exam that refer to specific tables that pertain to commercial calculations. The following table is one that you might see particularly pertaining to commercial installations. Refer to Table 6-3. This table would be used to determine the minimum size equipment grounding conductor for a raceway or a piece of equipment. The examples that follow help demonstrate what type of questions you may be asked concerning this table.

Table 6-3: Table 250.122: Minimum Size Equipment Grounding Conductors for Grounding Raceway and Equipment

Rating or Setting of Automatic Overcurrent Device in Circuit Ahead of Equipment, Conduit, Etc., Not Exceeding (Amperes)	Size (AWG or kcmil)	
	Copper	Aluminum or Copper-Clad Aluminum*
15	14	12
20	12	10
30	10	8
40	10	8
60	10	8
100	8	6

(continued on next page)

Table 6-3 *(continued)*

	Size (AWG or kcmil)	
Rating or Setting of Automatic Overcurrent Device in Circuit Ahead of Equipment, Conduit, Etc., Not Exceeding (Amperes)	**Copper**	**Aluminum or Copper-Clad Aluminum***
200	6	4
300	4	2
400	3	1
500	2	1/0
600	1	2/0
800	1/0	3/0
1000	2/0	4/0
1200	3/0	250
1600	4/0	350
2000	250	400
2500	350	600
3000	400	600
4000	500	800
5000	700	1200
6000	800	1200

Reprinted with permission from the NFPA 70-2005, *National Electrical Code*®, Copyright © 2004, National Fire Protection Association, Quincy, MA 02269. This reprinted material is not the complete and official position of the NFPA on the referenced subject, which is represented only by the standard in its entirety.

Example 1

What size copper equipment grounding conductor is needed for an overcurrent device rated at 40 amps?

The answer is a #10.

Example 2

What size aluminum equipment grounding conductor is needed for an overcurrent device rated at 200 amps?

The answer is a #4.

Summary

Much like Chapter 5, this chapter covers some popular questions concerning commercial load calculations. Some view commercial calculations as being easier than residential due to simpler calculations, but you should still approach them with caution. When performing such calculations, be sure to double-check your math as well as read each question carefully to ensure that you don't leave out any important details.

MOTOR CALCULATIONS

Motor calculations are often used on exams because of the numerous different scenarios that are available when installing motors. However, motor calculations can be difficult at times due to the number of steps that must be followed to find the correct solution. This unit will look at how to solve single-phase and three-phase motor problems, which range from simple to complex. Successfully performing motor calculations requires an understanding of some terms that will be described later as well as how to read some NEC table information. There are primarily six different calculations that can be performed and each will be discussed in detail.

Objectives

- Determine the full load current for single- and three-phase motors
- Size the motor overload protection device
- Calculate the correct branch circuit overcurrent device size
- Determine how to size the feeder conductor wire size
- Calculate the feeder overcurrent protection device

Question Types

The key terms listed throughout the chapter actually represent the six different types of questions that could be asked on an exam. For example, the exam may ask you to size the **branch circuit conductor** that serves one motor. Article 430.22 provides the information for determining the branch circuit conductor size but you will have to know how to determine the **full load current (FLC)** before completing the problem. Hint: Most motor calculation questions involve determining the FLC first before proceeding with the question. So, it is important to understand all of the aspects of motor terms and table information. Refer to Figure 7-1 to identify the different motor calculation components.

Figure 7-1

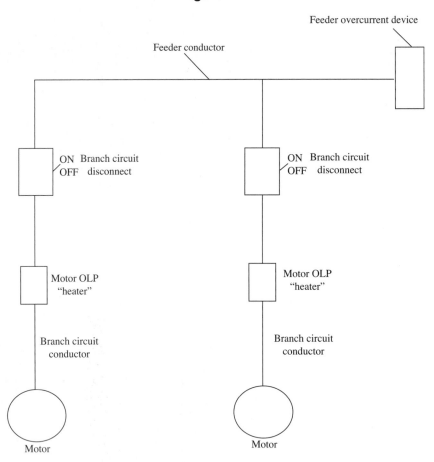

Branch circuit conductor — Article 430.22: the conductor that services a single motor. Found by multiplying FLC by 125%, then using Table 310.16 to find the proper size conductor.

Full load current (FLC) — Table 430.248 for single-phase motors and Table 430.250 for three-phase motors. Match the voltage and the horsepower to find the amperage.

Some state exams contain numerous motor questions because there are so many different types of questions that can be asked. They can also be some of the most difficult and therefore make excellent test questions. Calculations #1 through #4 pertain to single-motor applications whereas calculations #5 and #6 may contain questions with more than one motor on a single feeder.

Calculation #1: Finding the Full Load Current (FLC): Tables 430.248 and 430.250

When you are trying to determine the full load current or FLC, the motor nameplate is very useful. Most nameplates include the motor model number, the supply voltage options, horsepower rating, hertz, whether it is single-phase or three-phase, and other valuable information. To find the FLC, simply locate the horsepower rating and the voltage supplied to the motor from the nameplate information. Then, refer to Table 430.248 for single-phase motors and Table 430.250 for three-phase motors. Where the horsepower and voltage intersect, you will find the FLC. On most exams a picture of the nameplate will not be provided but the horsepower and voltage will be stated in the problem. Take a look at the sample motor nameplate in Figure 7-2.

Figure 7-2

Big Al's Motors, Inc.		
M00. AK2215	HP 1/2	FLA 7.6/3.8
V. 115/230	RPM 1725	Hz 60
PH 1	S.F. 125	KVA Code M
MTR No. KBS1619KDB Prelubricated bearings		To reserve rotation, interchange black and read leads

How to Use Tables 430.248 and 430.250

Use the horsepower rating and the different voltages provided in conjunction with Table 7-1 to determine the answer. First, determine if the motor is single phase or three phase. Looking at Figure 7-2, you can see that this example nameplate lists the motor as single phase. Then, locate the motor voltages on the motor nameplate information. The nameplate in Figure 7-2 is marked with 115/230 volts. This particular motor can be wired for 115 volts or 230 volts. Next, locate the horsepower rating on the nameplate. This motor is rated at ½ horsepower. Finally, use these three pieces of information to find the full load current (FLC). If the motor is single phase, you will use Table 430.248 to locate the FLC that is the intersection of the horsepower and the voltage rating. If the motor is three phase, use Table 430.250 to determine the FLC. Take a look at the example that follows to see how these two tables are used.

Table 7-1: Table 430.248: Full-Load Currents in Amperes, Single-Phase Alternating-Current Motors

The following values of full-load currents are for motors running at usual speeds and motors with normal torque characteristics. The voltages listed are rated motor voltages. The currents listed shall be permitted for system voltages ranges of 110 to 120 and 220 to 240 volts.

Horsepower	115 Volts	200 Volts	208 Volts	230 Volts
⅙	4.4	2.5	2.4	2.2
¼	5.8	3.3	3.2	2.9
⅓	7.2	4.1	4.0	3.6
½	9.8	5.6	5.4	4.9
¾	13.8	7.9	7.6	6.9

(continued on next page)

Table 7-1 *(continued)*

Horsepower	115 Volts	200 Volts	208 Volts	230 Volts
1	16	9.2	8.8	8.0
1 ½	20	11.5	11.0	10
2	24	13.8	13.2	12
3	34	19.6	18.7	17
5	56	32.2	30.8	28
7 ½	80	46.0	44.0	40
10	100	57.5	55.0	50

Reprinted with permission from the NFPA 70-2005, *National Electrical Code®*, Copyright © 2004, National Fire Protection Association, Quincy, MA 02269. This reprinted material is not the complete and official position of the NFPA on the referenced subject, which is represented only by the standard in its entirety.

Example 1

What are the full load currents for a motor based on the nameplate in Figure 7-2?

> At 115 Volts = **9.8 amps**
> At 230 Volts = **4.9 amps**

Example 2

What is the FLC for a single-phase, 208-volt, ½ hp motor?

> Table 430.248 = **5.4 amps**

Example 3

What is the FLC for a single-phase, 115-volt, 3 hp motor?

> Table 430.248 = **34 amps**

Example 4

What is the FLC for a single-phase, 240-volt, 1 hp motor?

> Table 430.248 = **8 amps**

If the question should ask about a three-phase motor, you would apply the same strategy for finding the FLC except you would use Table 430.250. The following examples demonstrate how to find the FLC for three-phase motors.

Example 5

What is the FLC for a three-phase, 208-volt, 10 hp motor?

> Table 430.250 = **30.8 amps**

Example 6

What is the FLC for a three-phase, 460-volt, 5 hp motor?

> Table 430.250 = **7.6 amps**

Calculation #2: Finding the Motor Overload Protection or "Heater" Size: Article 430.32

An exam question may ask you to find the proper size **motor overload protection** for a particular motor. Some motor nameplates may be marked with the amperage rating. If so, you should use the amperage rating listed on the nameplate to determine the size of the motor overload protection. However, if the nameplate does not contain this information, you will have to determine the size of the motor overload protection by using 430.32. There are two different code references that address this type of question. First, the minimum size is described in 430.32(A)(1). Second, the maximum size is found in 430.32(C). However, unless the question specifically asks for the maximum size, always use the minimum size information found in 430.32(A)(1). This section provides three categories to find the proper size: motors marked with a service factor, motors marked with a temperature rise, and all other motors. Figure 7-3 provides a look at a typical set of motor overload protection coils or what is often referred to in the field as "heaters."

Motor overload protection or (heater) — Article 430.32(A1) and 430.32(C): The thermal protection that protects a single motor

Figure 7-3

In the examples that follow, it is assumed that no nameplate amperage rating exists, so the motor overload protection should be sized by using the FLC. But remember, the first thing you should do is determine if the nameplate states the size of the overloads first and, if so, use that information to size the overloads.

Example 1

What size motor overload protection should be used for a 3 hp, 208-volt, single-phase motor?

> Since this question does not state a temperature or service factor, you should use the "all other motors" percentage of 115%.

Step 1. Table 430.248 = 18.7 FLC

Step 2. $18.7 \times 115\% = \textbf{21.5 amps}$

Example 2

What is the minimum size motor overload protection for a 5 hp, 230-volt, single-phase motor with a service factor of 1.3?

This question states a service factor of 1.3, which means that 125% must be used.

Step 1. Table 430.248 = 28 FLC
Step 2. $28 \times 125\% = $ **35 amps**

Calculation # 3: Finding the Branch Circuit Conductor Size: Article 430.22 and Table 310.16

This is the conductor that supplies a single motor. Some motor circuits may not be fed by a feeder circuit like the two motors in Figure 7-1. However, this is the conductor that originates from either the panel or the feeder circuit that supplies other motors. Article 430.22(A) states that the branch circuit conductors that supply a motor must not be less than 125% of the motor's FLC.

Example 1

What is the ampacity of the branch circuit conductor supplying a ½ hp, 115-volt, single- phase motor?

Step 1. Table 430.248 = 9.8 (FLC)
Step 2. 430.22(A) $9.8 \times 125\% = $ **12.25 amps**

Example 2

What size of THHN conductor should be used for a 1½ hp, 115-volt, single-phase motor?

Step 1. Table 430.248 = 20 amps (FLC)
Step 2. 430.22(A) $20 \times 125\% = 25$ amps
Step 3. Table 310.16 **#14**

Note: The * next to 14 in the table does not apply to conductor sizing but to overcurrent protection; therefore #14 is large enough.

Calculation #4: Finding the Branch Circuit Overcurrent Device: Table 430.52

Figure 7-4 shows a typical **branch circuit overcurrent device**. This calculation will determine the proper size fuses that need to be installed in order to effectively protect the circuit. Table 430.52 provides the information needed to find the branch circuit overcurrent protection device. Note that the type of motor and the type of fuse or breaker is used to determine the proper size. Also note that the word *polyphase* refers to three-phase motors.

Branch circuit overcurrent device — Table 430.52: The type of motor and the type of overcurrent device is used to determine the proper size. Article 240.6 lists the standard-sized devices. If the amperage size calculated does not match one of the devices listed, the next highest device should be chosen.

Figure 7-4

Table 7-2: Table 430.52: Maximum Rating or Setting of Motor Branch-Circuit Short-Circuit and Ground-Fault Protective Devices

Type of Motor	Percentages of Full-Load Current			
	Nontime Delay Fuse	Dual Element (Time-Delay) Fuse	Instantaneous Trip Breaker	Inverse Time Breaker
Single-phase motors	300	175	800	250
AC polyphase motors other than wound-rotor				
Squirrel cage-other than Design B energy-efficient	300	175	800	250
Design B energy-efficient	300	175	1100	250
Synchronous	300	175	800	250
Wound rotor	150	150	800	150
Direct current (constant voltage)	150	150	250	150

Notice that Table 7-2 lists a number of different types of motors. Under most circumstances, you normally would use only the single-phase motors and AC polyphase motors. You would refer to the other types of motors only if specifically called for in the exam question.

Example 1

What size dual element fuse should be used for a single-phase motor with an FLC of 12 amps?

FLC of 12 amps $\times$ 175% = 21 amps

Refer to 240.6(A); since there is no 21-amp fuse size you must select the next highest rating of **25 amps**.

Example 2

What size inverse time breaker should be used for a three-phase, 1 hp, 208-volt motor?

Note: Here you must first determine the FLC from Table 430.250

Table 430.250: FLC = 4.6 amps
4.6 amps $\times$ 250% = 11.5 amps
Refer to 240.6(A); the next highest breaker size is **15 amps**.

Calculation #5: Finding the Feeder Conductor Size: Article 430.24

Article 430.24 states that when conductors are supplying two or more motors, the ampacity should be equal to the sum of the full load current rating of all the motors plus 25% of the highest-rated motor on the feeder. It is important to understand that the largest motor may not be the motor with the greatest horsepower rating. The largest motor should be the motor that draws the most amps. For example, a 1 hp, 115-volt, single- phase motor has an FLC of 16 amps. However, a 2 hp, 230-volt, single-phase motor has an FLC of only 12 amps. Therefore, the 2 hp motor is actually smaller than the 1 hp motor with respect to ampacity. Also, remember that this calculation considers the ampacity of each phase conductor.

Example 1

What is the required **feeder conductor** ampacity for three 1 hp, 230-volt, single-phase motors?

First, find the FLC of the motors: Table 430.248 = 8 amps each

Second, add the three motor FLCs together: 8 + 8 + 8 = 24 amps, but you must take the largest of the three and add 25%,

Third, since all the motors are the same size, you can use any of the three to add the extra 25%.
So, 8 $\times$ 125% = 10 + 8 + 8 = **26 amps**

More simply put, (8 $\times$ 125%) + 8 + 8 =   (10) + 8 + 8 = **26 amps**

Feeder conductor — Article 430.24: Conductor calculated when more than one motor is used. The largest FLC is multiplied by 125% and each additional motor FLC is added at 100%.

Example 2

What is the required feeder conductor ampacity for a 2 hp, 115-volt, single-phase motor and a 1½ hp, 115-volt, single-phase motor?

FLC of 2 hp motor = 24 amps Table 430.248
FLC of 1½ hp motor = 20 amps Table 430.248
The 2 hp motor is larger than the 1½ hp motor so,

(24 $\times$ 125%) + 20 =   (30) + 20 = **50 amps**

Motor calculation questions can be very tricky and you may want to skip over some of them in the beginning and return to them after you answer some of the less challenging questions. One of tricky motor questions involves three-phase motors. When dealing with single-phase motors, you would have one lead of the motor connected to L1 and the other lead connected to L2. However, three-phase circuits present a more complicated problem for performing calculations. Three phase means you have phase A, phase B, phase C, and the neutral (N), or four conductors. One the three leads from the motor connects to a different phase.

If you were asked to calculate the feeder conductor size for two three-phase motors, you would perform the calculation as one having two motors on the feeder. However, think about a problem stating that one three-phase and three single- phase motors are on the same feeder. How many motors are on each phase? If you answer four you are wrong. The answer is actually that only two motors connected per phase. Phases A, B, C, and N are all used for the three-phase motor once each, and each phase is used once more for each single-phase motor. Therefore, the three-phase motor is connected to A, B, C, and N; one single-phase motor is connected to A and N; one single-phase motor is connected to B and N; and the last single-phase motor is connected to C and N.

Example 3

What should the ampacity of a feeder conductor be for one 5 hp, three-phase, 208-volt motor, and three 1 hp, single-phase, 115-volt motors?

> FLC of 5 hp motor = 16.7 amps Table 430.250
> FLC of 1 hp motors = 16 amps Table 430.24
>
> The largest motor is the 5 hp, so $(16.7 \times 125\%) + 16 = \quad (20.88) + 16 = \mathbf{36.88}$

Notice that only one of the single phase motors is added to the calculation because there are three hot conductors in the circuit. The three phase motor connects to all three phases and the single phase motors would be connected to only one of the each phases. Therefore, each phase conductor would only have the three phase motor connected and one single phase motor connected to it.

Taken a step further, you could be asked to find the correct AWG size of the feeder conductor. So, what size THW conductor should be used in the example above?

> Using Table 310.16 = **#8**

Calculation #6: Finding the Feeder Overcurrent Protection Device Size: 430.62(A)

The sizing of the **feeder overcurrent protection** device should be based upon the largest branch circuit overcurrent device plus the full load currents of all the other motors. The feeder overcurrent device is the disconnect that is located between the individual branch circuit disconnects and the main power supply or panelboard.

Feeder overcurrent protection — Article 430.62(A): The largest branch circuit overcurrent device would be used with the addition of all other FLCs of each motor connected to that feeder. The next smallest size should be used when selecting overcurrent protection.

Example 1

What size dual element fuse should be used to size the feeder overcurrent protection for the following feeder that serves two motors on two separate branches?

> Motor #1: Uses a 20-amp breaker on the branch circuit and has an FLC of 10 amps.
>
> Motor #2: Uses a 30-amp breaker on the branch circuit and has an FLC of 15 amps.
>
> Find the largest branch circuit protection, which is the 30-amp breaker.
> Next, add the other motor's FLC, which is 10 amps: 30 + 10 = 40 amps
> Finally, refer to 240.6(A) for a standard size, which is **40**.

Example 2

What size dual element fuse should be used to size the feeder overcurrent protection for the following feeder that serves three motors on three separate branches?

Motor #1: Uses a 20-amp breaker on the branch circuit and has an FLC of 10 amps.

Motor #2: Uses a 30-amp breaker on the branch circuit and has an FLC of 15 amps.

Motor #3: Uses a 40-amp breaker on the branch circuit and has an FLC of 20 amps.

Find the largest branch circuit protection, which is the 40-amp breaker.
Next, add the other motors' FLCs, which are 10 amps and 15 amps:

$$40 + 10 + 15 = 65$$

Finally, refer to 240.6(A) for a standard size, which is **60.**
Notice that there is no standard 65 available, so you must choose the **next lowest size**. *Remember to round down for feeder sizes.*

Motor Practice Problems

1. What is the FLC of a single-phase, 240 V, A/C motor with a 2-horsepower rating?
   a. 13.2 A
   b. 12 A
   c. 10 A
   d. 6.8 A

2. A _____ size overload should be used to protect a three-phase, 240 V, 3 hp motor with a service factor of 1.2.
   a. 12 A
   b. 14.2 A
   c. 9.6 A
   d. 10 A

3. What size THHN conductor is required for a 5 hp, 240 V, single-phase motor?
   a. #14
   b. #12
   c. #10
   d. #8

4. A _____ branch circuit dual element fuse is required for a 3 hp, single-phase, 240 V motor.
   a. 20 A
   b. 25 A
   c. 30 A
   d. 35 A

5. What is the amp requirement on the feeder conductor for a 2 hp motor and a 3 hp motor, both 240 V and single-phase, on the same feeder?
   a. 35 A
   b. 40 A
   c. 30 A
   d. 45 A

6. What size feeder overcurrent protection should be used with three motors on 25-amp branch circuits each and with an FLC of 10 amps each?
   a. 60 A
   b. 40 A
   c. 45 A
   d. 80 A

Three-Phase Motor Practice Problems

1. What size dual element fuse is required to protect the branch circuit for a three-phase, 20 hp, 208 V motor?
   a. 100 A
   b. 110 A
   c. 120 A
   d. 130 A

2. What is the minimum size branch circuit conductor allowed for a 25 hp, 480 V, three-phase motor that requires using 75 degree insulation?
   a. #6
   b. #10
   c. #4
   d. #8

3. A _____ size feeder conductor is required for a 1 hp, 240 V, single-phase motor and a 3 hp, 208 V, three-phase motor. The connection is rated for 60 degrees.
   a. #8
   b. #10
   c. #12
   d. none of these

4. What is the size of the feeder overcurrent protection for three 5 hp, 208 V, three-phase motors? Each motor branch circuit is protected by a 30-amp dual element fuse.
   a. 70 A
   b. 60 A
   c. 100 A
   d. none of these

5. What is the minimum size heater that can be used for a three-phase, 7½ hp, 208 V motor with a service factor of 1.15?
   a. 30.25 A
   b. 27.83 A
   c. 24.2 A
   d. 32.26 A

Motor Connections

Some exams may evaluate how to connect motors for high voltage and low voltage operation. You may encounter a question that contains a diagram of some motor leads, such as a nine-lead connection. The question may ask you to demonstrate how to connect the motor in a high-voltage delta configuration or how to connect the motor in a low- voltage wye configuration. The figures that follow demonstrate how to make low-voltage and high-voltage connects for both delta and wye connections. Figure 7-5 illustrates a high-voltage delta connection. Figure 7-6 illustrates a low-voltage delta connection. Figure 7-7 illustrates a high-voltage wye connection. Finally, Figure 7-8 illustrates a low-voltage wye connection.

Figure 7-5

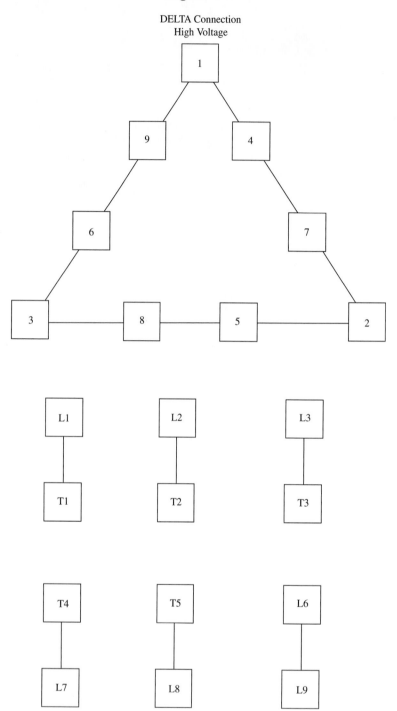

Figure 7-6

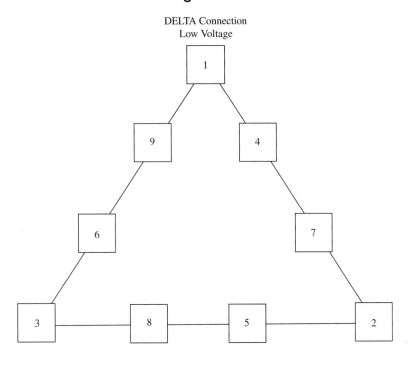

DELTA Connection
Low Voltage

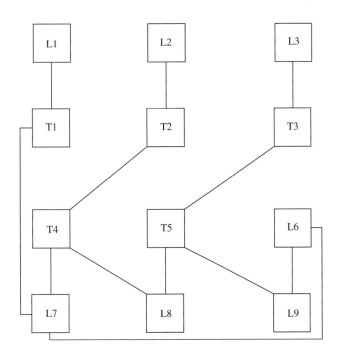

Figure 7-7

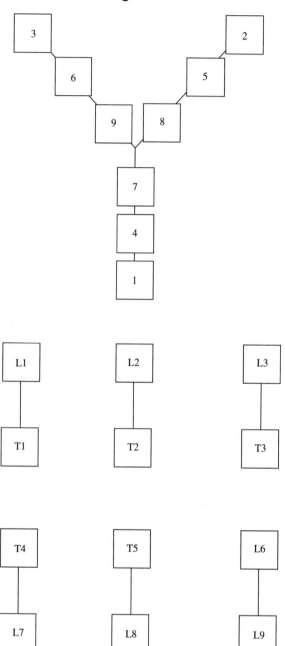

Figure 7-8

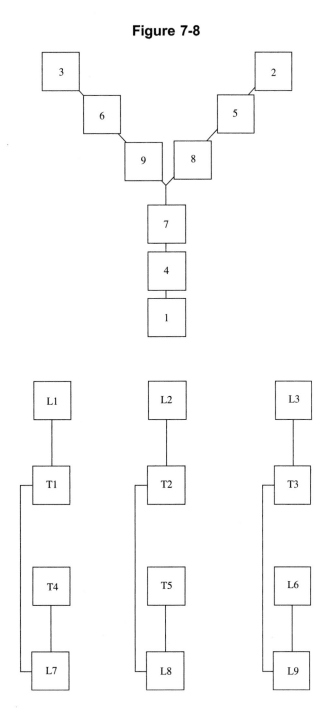

Summary

This chapter explored the many different types of motor calculations that you may experience on the exam. Motor calculation questions are very popular exam questions. They are on exams because they can be difficult and time consuming. When performing motor calculations, you must remember to read the question very carefully and to also recheck your math when you finish the problem. There may be some motor questions that require no calculations but rather ask you to demonstrate proper voltage connections, so be sure to be familiar with low-voltage and high-voltage connections. Some exams will have more motor questions than others, but you should feel confident that you will see some motor questions on almost every exam.

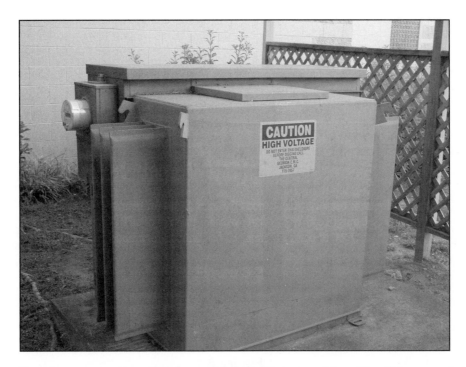

Transformers are often referred to as the most efficient machines. Very little energy is lost between the windings, making their use all the more important. There is a common rule that applies to transformers: the values of a transformer are proportional to its turns ratio. The turns ratio is the most important part of transformer calculations. Another important concept of transformers is that the power input (VA) is equal to the power output (VA). In other words, if a transformer has 280 VA input it must have 280 VA output. This chapter explores some of the many aspects of transformers and transformer calculations.

Objectives

- Define step-up and step-down transformers
- Discuss the rules concerning the turns ratio
- Define the function of the neutral
- Perform single-phase and three-phase transformer calculations

Basic Transformer Calculations

The following are some of the basic transformer formulas that you may use to determine transformer values. These formulas utilize cross multiplication and basic algebra techniques to solve the equations.

Ep = primary voltage   Es = secondary voltage

Np = number of turns in primary   Ns = number of turns in secondary

Ip = primary current   Is = secondary current

$$\frac{Ep}{Es} = \frac{Np}{Ns} \qquad \frac{Np}{Ns} = \frac{Is}{Ip} \qquad \frac{Ep}{Es} = \frac{Is}{Ip} \qquad Ep \times Ns = Es \times Np \qquad Np \times Ip = Ns \times Is$$

$$Ep \times Ip = Es \times Is$$

Turns Ratio

The turns ratio is one of the best methods to determine transformers' values. It is one of the simplest methods because all the values of a transformer are based on a ratio. For example, if there are twice the number of turns of wire on the **secondary winding** than the **primary winding**, then there must be twice the voltage on the secondary than on the primary. However, remember that current reacts the opposite of voltage in a transformer. For example, if the voltage in a certain transformer is four times greater in the secondary than the primary, the current will be four times less in the secondary than the primary. This will be demonstrated later in this chapter. By determining the ratio first, you should be able to complete the calculations more quickly. Look at Figure 8-1. Use the turns ratio to determine the values.

Secondary windings — The load side of the transformer or output

Primary windings — The power supply side of the transformer or input

Figure 8-1

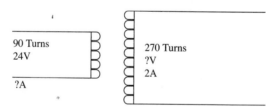

This is a **step-up transformer** because there are more turns on the secondary than the primary.

270 / 90 = 3   **The ratio is 1 : 3**

Step-up transformers — Transformers that contain more turns of wire on the secondary windings than on the primary windings. Therefore, the secondary voltage is higher than the primary voltage. An example of this would be the large transformers that step up the voltage at power plants. See Figure 8-2.

Figure 8-2

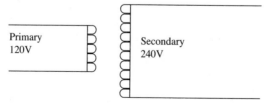

Rule: Voltage is directly proportional to the turns ratio. In other words, if the number of turns is increased by three times, then the voltage must increase by three times.

Since voltage is directly proportional to the ratio, simply multiply 24 V × 3 = **72 V** on the secondary.

In the previous example, what would be the amperage in the primary if the secondary has 2 amps?

Rule: Amperage is inversely proportional to the turns ratio. This means that amperage reacts opposite of voltage in relation to the ratio. In this example, the secondary is increased three times so the amperage must be decreased three times.

Therefore, if there are 2 amps in the secondary you would divide 2 by 3. 2 / 3 = **.66 A**

Neutral

Knowing how to calculate the neutral of a transformer is very important. The neutral carries the unbalanced current in the circuit. If only two wires are used in a circuit, then there is no true neutral present. There must be three wires to have a true neutral. The amount of current the neutral carries is the difference between the L1 current and L2 current. Refer to Figure 8-3.

Figure 8-3

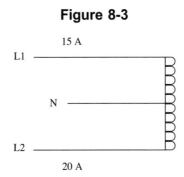

If L1 and L2 both carried 20 amps each, then the neutral would carry 0 amps.

Example 1

In Figure 8-3, how much could the neutral potentially carry?

If L1 were to open, then the neutral would have to carry all of L2, or 20 amps.

If L2 were to open, then the neutral would have to carry all of L1, or 15 amps.

So the maximum would be **20 amps.**

It is important that when you size the neutral, you should choose the correct size that would safely carry the maximum current in case one of the lines (L1 or L2) is opened.

Transformer Connections

Transformers can be connected in either series or in parallel depending upon the desired output. Transformer leads should be marked to indicate polarity and identify the individual leads, whether they be H1, H2, X1, X2, and so forth. Polarity dots are often used to assist in connecting the transformer to obtain the desired configuration. Some exams may simply use a diagram like the one in Figure 8-4 and ask whether the transformer is connected to indicate additive or subtractive polarity. In Figure 8-4, the polarity dots are located on X2 and H1, which would result in subtractive polarity. In Figure 8-5, the polarity dots are located on X1 and H1, which would result in additive polarity.

Figure 8-4

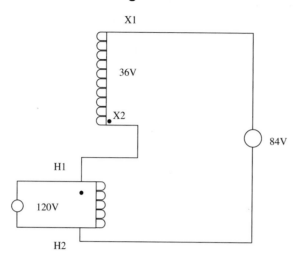

Figure 8-5

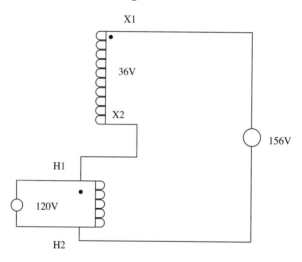

Series or Parallel Transformer Connections

Some exams may include diagrams like the ones in Figure 8-6. Here are two figures that represent series transformer connections and parallel transformer connections. The exam may simply ask you to identify which configuration is indicated in the diagram. Notice that the first transformer in Figure 8-6 is connected in series and the second transformer is connected in parallel. In other words, the same transformer can be connected to produce 240 volts on the secondary or 120 volts on the secondary simply by wiring the secondary in series or in parallel with the primary.

Figure 8-6

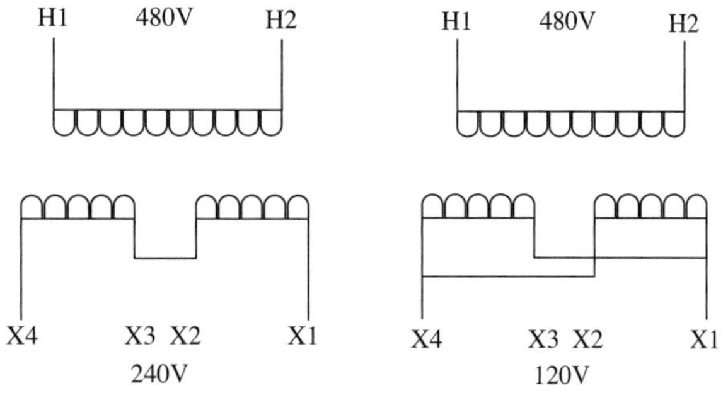

H1   480V   H2   H1   480V   H2

X4   X3 X2   X1   X4   X3 X2   X1
       240V   120V

More Transformer Calculations

A transformer has 480 volts on the primary, 120 turns on the primary, 80 turns on the secondary. What is the secondary voltage?

First, find the correct formula where at least three of the values are known.

The turns ratio = 120 / 80 or 1.5 : 1

$$\frac{Ep}{Es} = \frac{Np}{Ns} \qquad \frac{480}{Es} = \frac{120}{80}$$ Now, cross multiply. $480 \times 80 = 38,400$

$$\frac{38,400}{120} = \frac{120Es}{120}$$

$$Es = \textbf{320 volts}$$

However, an easier way to determine this answer would be to use the turns ratio.

120 turns on primary/80 turns on secondary = 1.5 or a **1.5:1 ratio**

There are more turns on the primary than on the secondary so this is a **step-down transformer**.
Simply, divide 480 by 1.5 $480 / 1.5 = \textbf{320 volts}$

Step-down transformers — Transformers that contain more turns of wire on the primary windings than on the secondary windings. Therefore, the primary voltage is higher than the secondary voltage. An example of this would be the transformer that is located on the utility pole at your home. The supply voltage is stepped down to 240 volts to serve your home. See Figure 8–7.

Figure 8-7

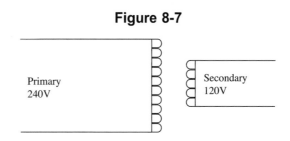

Primary
240V

Secondary
120V

Example 1

The turns ratio for the transformer in Figure 8-8 is:

$480 / 120 = 4$ or **1 : 4 ratio**

Figure 8-8

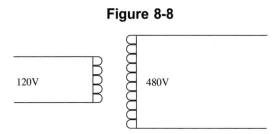

Example 2

If the input of the transformer above in Figure 8-8 is 1,200 VA, what is the output?

Remember, input (VA) also equals output (VA). So, output must equal **1,200 VA**.

Transformer Calculation Practice Problems

1. What is the primary current for the transformer in Figure 8-9?
    a. 80 A
    b. 20A
    c. 40 A
    d. none listed

Figure 8-9

240V
40A

480V
?A

2. How many amps does the neutral carry if L1 carries 25 amps and L2 carries 18 amps?
    a. 10 A
    b. 43 A
    c. 7 A
    d. 6 A

3. A transformer has 480 V on the primary, 10 amps on the primary, and 40 amps on the secondary. What is the secondary voltage?
    a. 240 V
    b. 208 V
    c. 120 V
    d. 480 V

4. In question 3, if the output is 4,800 VA, what is the input?
    a. 4,800 VA
    b. 400 VA
    c. 4,000 VA
    d. 2,800 VA

Three-Phase Transformers

The symbols in Figure 8-10 demonstrate the two different three-phase transformer configurations. Three-phase calculations are conducted differently than with single-phase transformers. The line and phase values have to be considered and the square root of three must be factored in as well. The formulas that follow help demonstrate how to properly determine three-phase values.

Figure 8-10

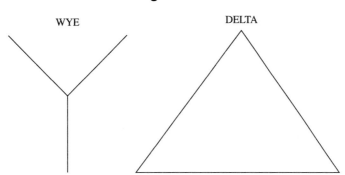

Wye: E line = E phase × 1.732   E phase = E line / 1.732

   I line = I phase

Delta: I line = I phase × 1.732   I line = I phase / 1.732

   E line = E phase

Three-phase power: VA = 1.732 × E line × I line   (When line values are known)
Three-phase power: VA = 3 × E phase × I phase   (When phase values are known)

Example 1

A wye-connected transformer has a line voltage of 208 volts. What is the phase voltage?

   E phase = E line / 1.732
   E phase = 208 / 1.732 = **120 volts**

Example 2

A delta-connected transformer has a line voltage of 480 volts. What is the phase voltage?

   E line = E phase

   Therefore, if E line is the same as E phase, the answer would be **480 volts**.

Three-Phase Practice Problems

1. A wye-connected transformer has a line current of 20 amps. What is the phase current?
   a. 10 A
   b. 40 A
   c. 34.64 A
   d. 20 A

2. A wye-connected transformer has a phase voltage of 208 volts. What is the line voltage?
   a. 360 V
   b. 208 V
   c. 480 V
   d. 220 V

3. A delta-connected transformer has a line current of 60 A. What is the phase current?
   a. 24.6 A
   b. 60 A
   c. 34.6 A
   d. 120.8 A

4. What is the volt-amp measurement for a transformer with a line voltage of 240 volts and a line current of 200 amps?
   a. 78,960 VA
   b. 83,136 VA
   c. 54,678 VA
   d. none listed

5. A delta-connected transformer has a phase voltage of 240 volts. What is the line voltage?
   a. 240 V
   b. 208 V
   c. 120 V
   d. 480 V

6. What is the three-phase power for a transformer with a phase voltage of 208 volts and a phase current of 35 amps?
   a. 12,609 VA
   b. 10,560 VA
   c. 7,280 VA
   d. 21,840 VA

Summary

    In this chapter, the basic operation as well as the basic types of transformers were described. Some exams will contain a number of transformer calculations while others may contain only a few. If your exam does contain several of these types of questions, you may want to skip the more difficult transformer questions and return to them if time allows, due to the fact that they can become complex and require a lot of your valuable time. As you can see, it is easy to confuse the different types of transformers when performing a calculation, so pay extra attention to the questions concerning transformers.

MOTOR CONTROLS

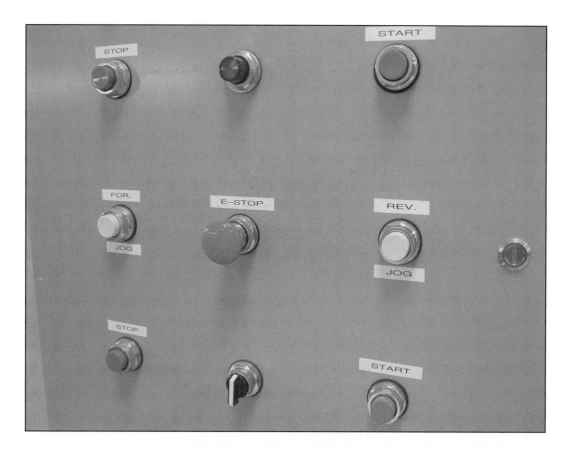

On most exams, the questions concerning motor controls are usually dealing with connections and operation questions. This is not typically a large section of the exam, but you could encounter some motor control questions. This chapter covers some basic motor control line diagrams and various component identifications. Some exams may actually contain line diagrams like the ones used in this chapter, while others may only refer to motor control circuits in reference to sizing and installation. However, it doesn't hurt to review some basic motor controls to be well prepared for whatever the exam may throw at you.

Objectives

- Identify basic motor control symbols
- Define and describe the different types of motor control logic
- Discuss the function and operation of line diagrams

Motor Control Symbols

Figure 9-1 demonstrates various basic motor control symbols. Some exam questions may simply ask you to identify different types of symbols. The ones shown in Figure 9-1 do not include all the symbols used on line diagrams but they do provide a good overview of what you might see on the exam.

Figure 9-1

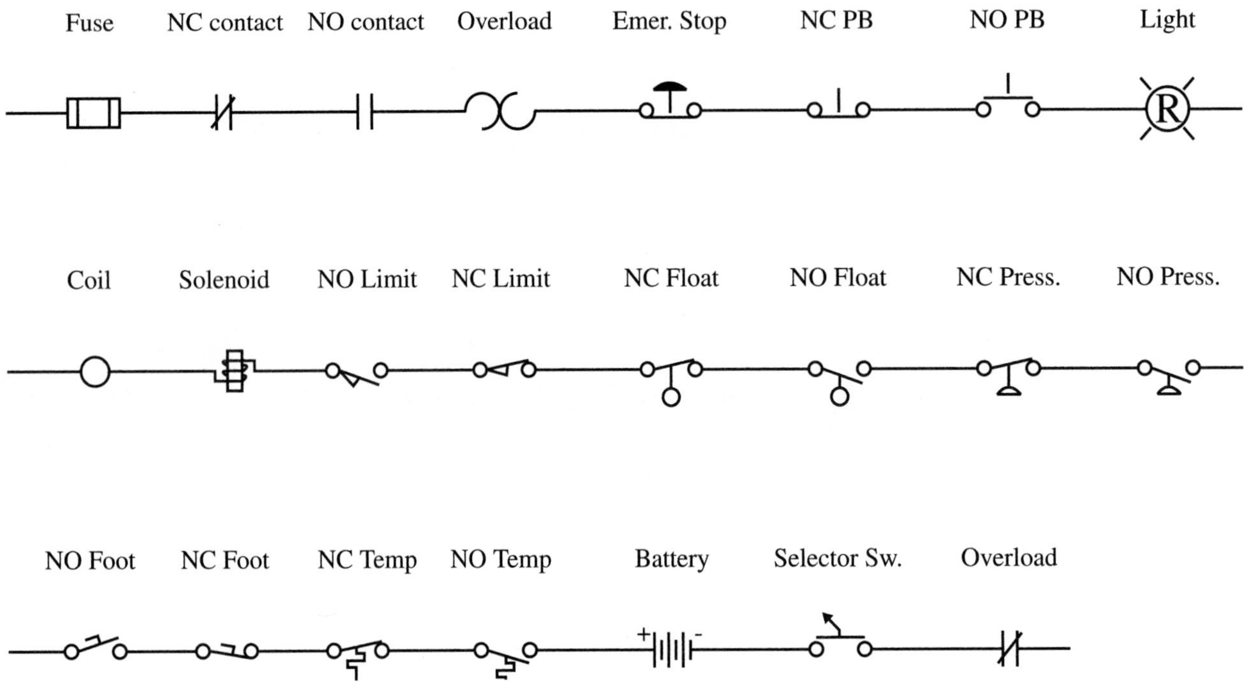

Fuse	NC contact	NO contact	Overload	Emer. Stop	NC PB	NO PB	Light

Coil	Solenoid	NO Limit	NC Limit	NC Float	NO Float	NC Press.	NO Press.

NO Foot	NC Foot	NC Temp	NO Temp	Battery	Selector Sw.	Overload

Line Diagrams

Line diagrams serve as sort of a road map to applying and understanding motor control circuits. Some exams may include such diagrams and ask you to determine the operation of a line diagram. Therefore, it is worth some time to take a look at some basic line diagrams and gain an overview of what might appear on the exam. Figure 9-2 shows a basic line diagram. It includes a stop button, start button, and light. The stop button is normally closed (NC) and the start button is normally open (NO). Pressing the start button will turn the light on.

Figure 9-2

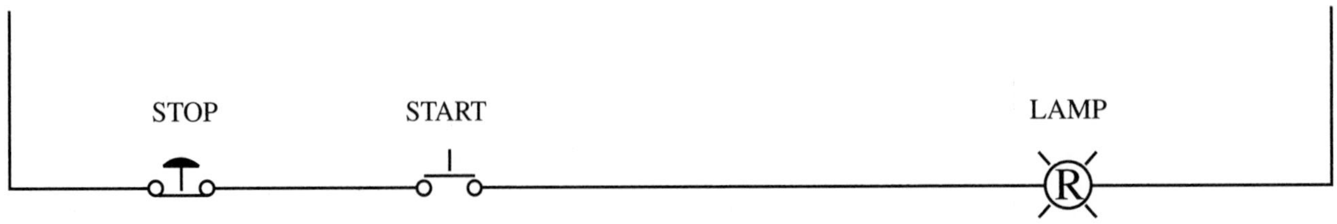

STOP   START   LAMP

Motor Control Logic

Developing an understanding of motor control logic is very important in order to be able to understand how line diagrams are developed and how they should be read. Logic functions make up how a line diagram is designed and used to control a specific operation or perform a certain function. There are several different forms of logic, and they can be combined to form other types of logic. The basic logic functions discussed in this chapter are AND, OR, AND/OR, NOT, and NOR. The figures that follow will help you see how each logic function operates in the line diagram. An exam question may ask you to identify different types of logic functions.

AND Logic

AND logic is created by using two or more normally open contacts, connected in series, to control an output or load. In Figure 9-3, you can see how PB1 and PB2 are connected in series. Both PB1 and PB2 must be pressed in order for the indicator light to come on. See Figure 9-3.

Figure 9-3

OR Logic

OR logic is accomplished when two normally open contacts are connected in parallel. In Figure 9-4, you can see how PB1 and PB2 are connected in parallel. In order to turn the light on, you can press either PB1 or PB2. See Figure 9-4.

Figure 9-4

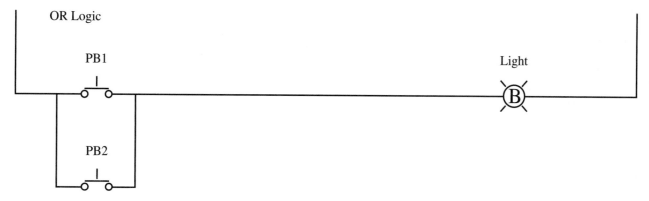

AND/OR Combination

This logic function is the combination of AND logic and OR logic. As you can see in Figure 9-5, PB1 and PB2 make up the AND logic and PB3 and PB4 utilize OR logic. In order to turn on the light in this circuit, you must press PB1, PB2, and either PB3 or PB4. See Figure 9-5.

Figure 9-5

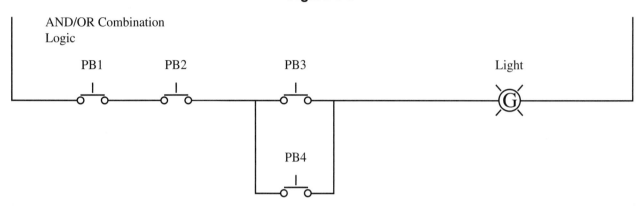

NOT Logic

This form of logic is accomplished by using a normally closed set of contacts. In this case, the load is on until the normally closed set of contacts are pressed open. An example of this would be the light in a refrigerator. When the door on a refrigerator is closed, the normally closed contacts are held open by the door and therefore the light is off. When the door is opened, the normally closed contacts spring back to the closed position and the light turns on. See Figure 9-6.

Figure 9-6

NOR Logic

NOR logic is derived from NOT logic. The difference between NOT and NOR is that NOR logic uses two or more normally closed contacts in series to operate a load. See Figure 9-7

Figure 9-7

Motor Control Practice Questions

Refer to Figure 9-8 for questions 1–3.

Figure 9-8

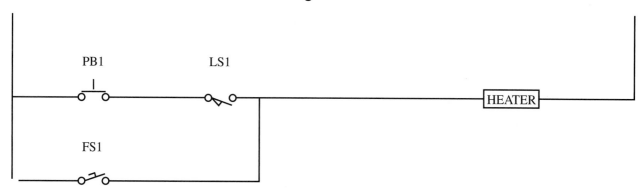

1. Under what conditions would the heater operate?
   a. PB1 is pressed
   b. FS1 is pressed
   c. Both PB1 and LS1 are pressed
   d. Both B and C are pressed

2. LS1 is a _____ switch.
   a. temperature
   b. limit
   c. safety
   d. float

3. FS1 is a(n) _____ switch.
   a. limit
   b. emergency stop
   c. pressure
   d. foot

Refer to Figure 9-9 for questions 4–6.

Figure 9-9

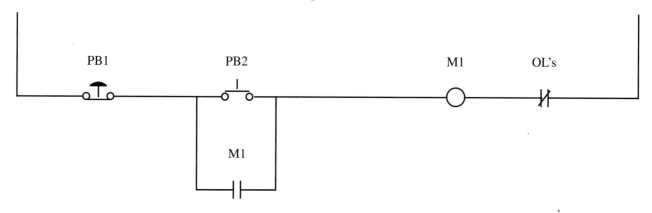

4. PB1 is a(n) _____ switch.
   a. emergency stop
   b. start
   c. run
   d. overload

5. Energizing coil M1 will close which set of contacts?
   a. start
   b. M1
   c. overload
   d. none listed

6. PB2 is a(n) _____ switch.
   a. NO
   b. NC
   c. stop
   d. jog

Refer to Figure 9-10 for questions 7–9.

Figure 9-10

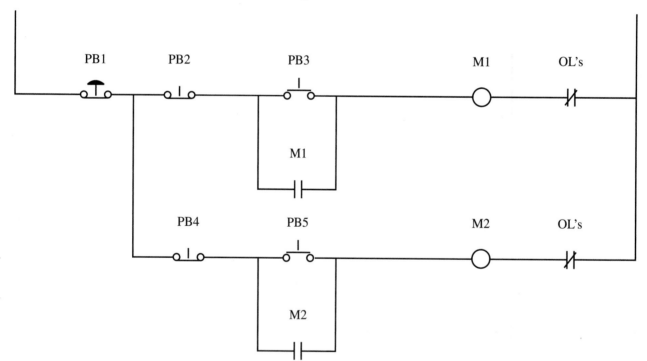

7. Pressing which button will activate M2?
   a. PB4
   b. PB5
   c. PB3
   d. PB1

8. PB4 is a(n) _____ push button.
   a. start
   b. NC
   c. locked
   d. NO

9. Once memory M2 has closed, which will return to an NO position?
   a. PB5
   b. PB3
   c. PB2
   d. PB4

More Line Diagram Examples

Here are some additional line diagram examples. Take a look at Figure 9-11. Here you can see that on line 1, there is an emergency stop followed by a stop/start station. Another start/stop station on line 3 is wired in parallel with the start/stop station located on line 1. As you following the diagram, you can see that when coil M1 located on line 1 closes, memory contacts M1 will close and contacts M1 located on line 5 will open. This will cause the pilot light located on line 5 to illuminate. When coil M2 on line 3 is energized, contacts M2 on line 4 will close and contacts M2 on line 6 will close. This will then energize the solenoid located on line 6.

Figure 9-11

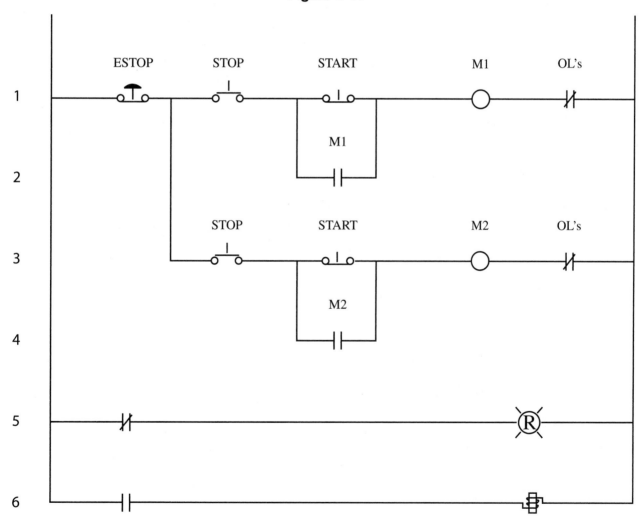

Line Diagram Practice Questions

Refer to Figure 9-12 for questions 10 and 11.

Figure 9-12

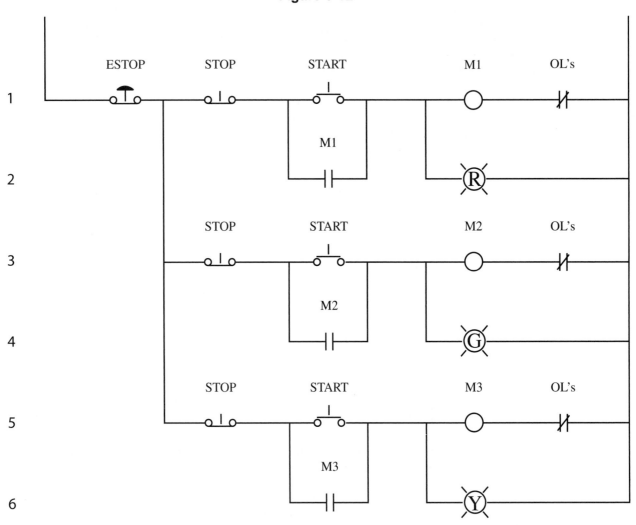

10. Coil M3 can be energized by _____ .
    a. pressing the start on line 1
    b. pressing the start on line 2
    c. pressing the start on line 3
    d. pressing any start button

11. The green pilot light will illuminate when _____ is energized and is wired in _____ with the load.
    a. M1, parallel
    b. M1, series
    c. M2, series
    d. M2, parallel

Summary

Chapter 9 covered some of the basic requirements for motor controls questions that you may experience on the exam. Some motor control questions may involve motor disconnect sizing, location, and connections that lie outside of the scope of this chapter. Chapter 9 attempts to make you aware of some different types of questions and the diagrams that you may see on the exam. Be sure to become familiar with various motor control symbols as well as Article 430 of the NEC. Article 430 covers not only motors but motor controllers and disconnecting means.

CHAPTER
10
SPECIAL EQUIPMENT AND LOCATIONS

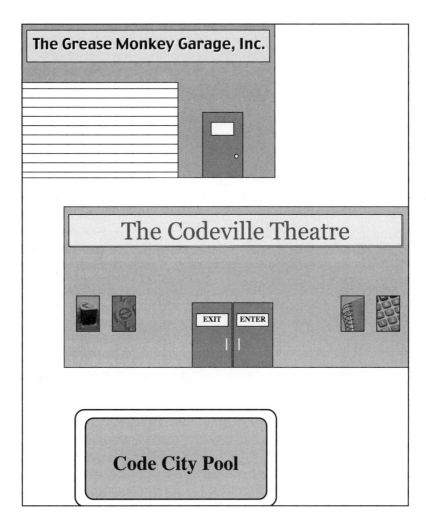

This chapter covers a couple of different installations including mobile homes and electric welders. There will probably not be a tremendous number of questions that refer to these items, but they are certainly worth mentioning. Some state exams ask questions that concern items such as mobile homes and welders because they make for good calculations. In addition to the items covered in this chapter, you may want to look at some other special locations like swimming pools and low-voltage applications before taking the exam.

Objectives

- Determine how to calculate mobile home service demands
- Discuss sizing for arc welders and resistance welders
- Discuss swimming pools and other bodies of water
- Discuss hazardous locations

Mobile Home Parks: Article 550

If the exam were to ask you to calculate the service demand for a mobile home park, you would refer to Article 550.31. The example below demonstrates how to calculate a basic load demand for 12 mobile homes. Table 10-1 lists the demand factors for a particular number of mobile homes. See Figure 10-1.

Figure 10-1

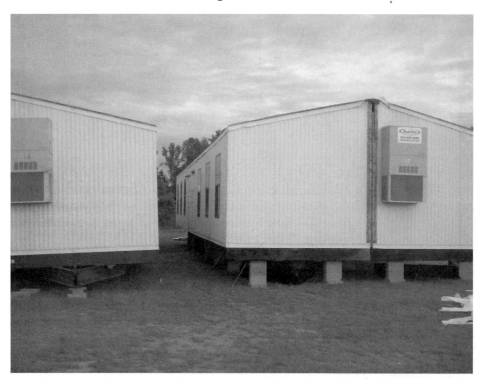

Example 1

Suppose you were asked to calculate the service demand for a mobile home park that is to contain 12 mobile homes.

First, refer to 550.31 (1) and (2). You must choose to use 16,000 VA or the largest calculated load for a particular mobile home. You could determine the demand for each mobile home by following the steps outlined in 550.18. The calculations for an individual mobile home are similar to a residential load calculation discussed in Chapter 5. However, since the information necessary to conduct the calculation for this example is not given in the above problem, you must use 16,000 VA. The problem would be worked as follows:

$$\frac{\text{VA for each} \times \text{\# of homes} \times \text{demand factor}}{240 \text{ V}} = \text{Load}$$

$$\frac{16,000 \text{ VA} \times 12 \times 27\%}{240 \text{ V}} = \frac{51,840}{240 \text{ V}} = \textbf{216 amps}$$

The 27% demand factor is found in Table 550.31.

Example 2

What is the service demand for a mobile home park containing 20 units with the largest unit having an individual demand of 19,000 VA?

Since 19,000 VA is larger than 16,000 VA, you should use 19,000 VA in the equation.

$$\frac{19,000 \text{ VA} \times 20 \times 25\%}{240 \text{ V}} = \frac{95,000}{240 \text{ V}} = \textbf{395 amps}$$

Table 10-1: Table 550.31: Demand Factors for Services and Feeders

Number of Mobile Homes	Demand Factors (percent)
1	100
2	55
3	44
4	39
5	33
6	29
7–9	28
10–12	27
13–15	26
16–21	25
22–40	24
41–60	23
61 and over	22

Reprinted with permission from the NFPA 70-2005, *National Electrical Code*®, Copyright © 2004, National Fire Protection Association, Quincy, MA 02269. This reprinted material is not the complete and official position of the NFPA on the referenced subject, which is represented only by the standard in its entirety.

Mobile Home Practice Problems

1. What is the service demand factor for a park containing 45 mobile homes?
   a. 29
   b. 25
   c. 22
   d. 23

2. What is the service demand amperage for a mobile home park with 15 mobile homes?
   a. 260 A
   b. 420 A
   c. 280 A
   d. none listed

3. What is the service demand amperage for a mobile home park with 90 units based upon a minimum of 20,000 VA for each unit?
   a. 1,250 A
   b. 1,800 A
   c. 1,650 A
   d. 2,125 A

4. All under-chassis wiring of a mobile home unit and where it is exposed to weather shall be installed in:
   a. EMT
   b. liquidtight
   c. PVC
   d. rigid or IMT conduit

5. The minimum service size allowed for a mobile home shall be not less than _____ amps.
   a. 75
   b. 90
   c. 100
   d. 125

Welders, Article 630

The NEC addresses two different types of welders: Arc welders, which include nonmotor generated and motor generated, and resistance welders. It is important to pay attention to which type of welder the question is referring to, as the calculations for each type are different. Welder questions can be a little tricky because they involve a lot of math, so be sure to work through the calculation slowly. Remember that the size of the supply conductors is based upon the duty cycle.

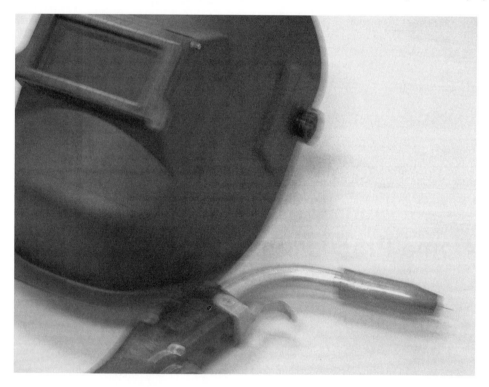

Arc Welders: 630.11– 630.15

If a question refers to an arc welder, you must also determine if it is talking about a nonmotor-generated or a motor-generated welder. Next, locate the duty cycle of the welder and multiply the current rating by the multiplier in Table 630.11(A).

Single Arc Welder, 630.11(A)

Example 1

A nonmotor-generated arc welder has a duty cycle of 70 with a rated primary current of 50 amps. The equipment is rated for 60°C. What is the required ampacity of the supply conductor?

Current rating × duty cycle multiplier = amps

$50 \times 0.84 = \textbf{42 amps}$

Conductor Size

The question could have gone a step further and asked you to find the correct size of conductor. If that were the case, you would refer to Table 310.16 for the wire size, which is a **#6** in the 60° column.

Overcurrent Protection

To size the overcurrent protection device you should refer to 630.12(A). It should be sized not more than 200% of the primary current of the welder. Take the primary current rating and multiply by 200%.

$50 \times 200\% = \textbf{100-amp breaker}$

Table 10-2: Table 630.11(A): Duty Cycle Multiplication Factors for Arc Welders

Duty Cycle	Multiplier for Arc Welders	
	Nonmotor Generator	Motor Generator
100	1.00	1.00
90	0.95	0.96
80	0.89	0.91
70	0.84	0.86
60	0.78	0.81
50	0.71	0.75
40	0.63	0.69
30	0.55	0.62
20 or less	0.45	0.55

Group of Arc Welders, 630.11(B)

Example 1

Four arc welders are to be connected to a circuit. All the welders are motor generators. What is the total amperage of the following welders?

Welder 1 – 70 A, 60% cycle
Welder 2 – 60 A, 60% cycle
Welder 3 – 60 A, 60% cycle
Welder 4 – 50 A, 60% cycle

#1	$70 \times 0.86 \times 100\% = 60.2$ amps
#2	$60 \times 0.81 \times 100\% = 48.6$ amps
#3	$60 \times 0.81 \times 85\% = 41.31$ amps
#4	$50 \times 0.75 \times 70\% = \underline{26.25 \text{ amps}}$

176.36 amps

Resistance Welders: 630.31 – 630.34

Resistance welders are calculated similarly to arc welders; however there are different multipliers and the method for a group of welders is different.

Single Resistance Welder: 630.31(A)

Example 1

A resistance welder has a duty cycle of 50% and a primary current of 60 amps. The equipment is rated for 60°C. What is the ampacity of this welder?

Current rating × duty cycle multiplier = amps

$60 \times 0.71 = \textbf{42.6 amps}$

Conductor Size

The question could have gone a step further and asked you to find the correct size of conductor. If that were the case, you would refer to Table 310.16 for the wire size, which is a **#6** in the 60° column.

Overcurrent Protection

To size the overcurrent protection device you should refer to 630.32(A). It should be sized not more than 300% of the primary current of the welder. Take the primary current rating and multiply by 300%.

Table 10-3: Table 630.31(A)(2) Duty Cycle Multiplication Factors For Resistance Welders

Duty Cycle (percent)	Multiplier
50	0.71
40	0.63
30	0.55
25	0.50
20	0.45
15	0.39

(continued on next page)

Table 10-3 *(continued)*

Duty Cycle (percent)	Multiplier
10	0.32
7.5	0.27
5 or less	0.22

Reprinted with permission from the NFPA 70-2005, *National Electrical Code*®, Copyright © 2004, National Fire Protection Association, Quincy, MA 02269. This reprinted material is not the complete and official position of the NFPA on the referenced subject, which is represented only by the standard in its entirety.

Group of Welders, 630.31(B)

Example 1

Three resistance welders are to be connected to a circuit. What is the total amperage for the following welders?

Welder #1 60 A, 40% cycle
Welder #2 50 A, 40% cycle
Welder #3 40 A, 40% cycle

#1 $60 \times 0.63 =$ 37.8 amps
#2 $50 \times 0.63 \times 60\% = 18.9$ amps
#3 $40 \times 0.63 \times 60\% = \underline{15.12 \text{ amps}}$

                    71.82 amps

Welder Practice Problems

1. What size overcurrent device is needed for a nonmotor generator arc welder with a primary current of 65 amps and a duty cycle of 70%?
   a. 110 A
   b. 90 A
   c. 55 A
   d. 60 A

2. What is the ampacity for a resistance welder with a primary current rating of 40 A and a duty cycle of 50%?
   a. 30 A
   b. 38.2 A
   c. 28.4 A
   d. 20 A

3. Refer to question 2. What would be the proper overcurrent protection for this welder?
   a. 120 A
   b. 125 A
   c. 135 A
   d. 140 A

4. Four 60-amp rated, 50% duty cycle motor generator arc welders are to be installed on a circuit. What is the amperage of the group of welders?
   a. 175.2 A
   b. 125 A
   c. 180 A
   d. 159.75 A

5. What is the ampacity for a nonmotor generator arc welder with a current rating of 75 amps and a duty cycle of 80%?
    a. 66.75 A
    b. 56.75A
    c. 65 A
    d. 76.5 A

6. Five resistance welders are installed in a local technical college. Determine the total ampacity of the five welders listed below.

    Welder #1   70 A, 50% duty cycle

    Welder #2   60 A, 50% duty cycle

    Welder #3   50 A, 50% duty cycle

    Welder #4   50 A, 50% duty cycle

    Welder #5   50 A, 50% duty cycle

    a. 124.6 A
    b. 139.16 A
    c. 148 A
    d. 136.2 A

Swimming Pools and Similar Installations: Article 680

Many exams use questions about swimming pools or other bodies of water. One of the first things you should do is become familiar with the definitions associated with pools or similar installations, and you can do this by referring to Article 680.2. Most of the questions concerning pools or other bodies of water will probably be basic NEC reference questions such as clearance or grounding, to name a couple. You may see some questions such as the following ones.

Example 1

What is the distance permitted for underground wiring located near a pool?

Not less than 5 feet from the inside wall of the pool unless the wiring is necessary for supplying the pool with power. Article 680.10

Example 2

What is the maximum length allowed for liquidtight flexible metal conduit used to connect a spa or hot tub?

Not more than 6 feet. Article 680.42(A)(1)

Hazardous Locations: Article 500

Another area of the NEC that you may see questions drawn from are hazardous locations, in Article 500. There are three hazardous location classes, and each class has two divisions. Hazardous locations are not limited just to Article 500, but more detail is provided for each class in Articles 501, 502, and 503. An exam could ask you a variety of question out of these sections of the NEC, such as the definition of each class, types of conduit installations allowed in each class, or grounding requirements, to name a few. You may see some questions like the ones that follow.

Example 1

Which hazardous class contains flammable gases or vapors?

Class I locations. Article 500.5(B)

Example 2

All boxes and fittings used in Class II locations shall be:

    a. raintight
    b. watertight
    c. dusttight
    d. explosion proof
    (C) dusttight. Article 502.10(B)(4)

Summary

Although you may not see a lot of questions concerning special equipment and locations on your exam, it never hurts to be prepared, since every exam is different. You need to be as familiar with every aspect of the NEC as possible because missing a couple of questions just because you are not familiar with the material may cost you a passing grade on the exam. If you have some spare time after you have thoroughly worked through this book, you may want to spend a little time reviewing other special conditions and locations such as electric signs, health care facilities, and other specific locations. References to other conditions and locations can be found in the second half of the NEC.

CHAPTER 11

BUSINESS LAW REVIEW

This portion of the book focuses on the business and tax section of an electrical contractor's exam. Not all exams include business and tax sections. However, it is advisable to gain as much knowledge as you can about business matters when conducting a contracting business, even if it is not included on the exam. Please note that certain sections of this business law portion require that you obtain additional information on federal taxes and employment laws, all of which can be found via the Internet.

Objectives

- Review basic business and accounting terms
- Discuss the different forms of business
- Demonstrate how to read and use financial statements
- Calculate key financial ratios for financial analysis

Reference Material

The reference material used in this book covers a number of important components of conducting business. Three primary references are used, and the following provides a brief description of each and where each can be located.

Circular E Tax Information

The tax information is based upon the 2006 Circular E, Employer's Tax Guide. Circular E discusses important information such as tax filing deadlines, recordkeeping, definitions of employees, wages/compensation, employee wage withholding, depositing, FUTA tax, and so forth. This document can be accessed through <http://www.irs.gov>.

Fair Labor Standards Act

The purpose of the FLSA is to set minimum wage, overtime, and child labor laws. This document demonstrates how to calculate wages and overtime pay. It also briefly describes enforcement and other important labor laws. Much of the FLSA can be found in the *Basic Business and Project Management for Contractors* book discussed below.

Business and Project Management for Contractors

One of the books allowed on some state electrical contractor's exams is *Basic Business and Project Management for Contractors,* which describes the different forms of business, business insurance, OSHA regulations, labor laws, **accounting** and financial systems, and contracts. The *Basic Business and Project Management for Contractors* is generic to all states. However, states like Alabama, Connecticut, Georgia, Louisiana, Maryland, North Carolina, West Virginia, and Virginia all have their own editions of this book, which covers specific information pertaining to that state's labor laws. For example, if you are take the exam in Georgia the title of the book would be *Business and Project Management for Contractors, Georgia Edition.* Also, some states may have an edition specifically for general business or for electrical contracting. Be sure and check with your state licensing board to see if this book pertains to your exam and if it is allowed in the testing center. The business information in this chapter is based upon *Basic Business and Project Management for Contractors,* 5th edition, which can be purchased at www.constructionbook.com (item #290-4556-04). This version provides a broad overview of information that pertains to all states.

Accounting — A method of organizing, recording, and using financial information to determine the financial status and forecast of a business

Accrual basis of accounting — System in which revenue is recorded when it is earned, not when cash is received, and expenses are recorded when incurred, not when cash is paid out

Cash basis of accounting — System in which revenue is recorded when cash is received even if it is collected in a different period than when it is earned

Note to the Reader

This chapter attempts to familiarize you with general business and accounting principles that you may or may not encounter on your electrical exam. You should note that not all state exams are the same, and you should consult your state authority concerning the material covered in your particular state. In addition, some of the tax calculations in this chapter may use tax information specific to a particular year and should be used only as an example of how to determine the proper method in tax calculation. You should consult the federal and state tax requirements that affect your state exam in addition to the material in this chapter. The chapter is limited in scope and does not include all the information needed to form and operate a business.

Forms of Businesses

Sole proprietorship: a business owned by a single person. Most businesses in the United States are sole proprietorships because they need little capital and are easy to start. However, it may not be the wisest form of businesses due to unlimited **liability**. If you are sued, you could stand to lose not only your business **assets** but also your personal assets. Many electrical contractors dream of owning their own sole proprietorship but don't understand the legal ramifications.

Liabilities — Debts owed to others

Assets — Tangible or intangible things that are owned

Partnership: a business owned by two or more persons. The advantage of a partnership is that capital and knowledge can be shared to form a stronger business. There are two basic forms of partnerships, general partnerships and limited partnerships. General partnerships are usually not a good idea because each partner is liable for debts accrued by the other partner(s). A limited partnership is made up of general partner(s) and limited partner(s). The disadvantage is that the general partner is still personally liable for the company's debt and that the limited partners, although they enjoy limited liability, are restricted in their control over the company.

Corporation: a business that is a separate legal entity and is owned by shareholders. The best way to protect your personal assets is to form a corporation. This will limit liability to the separate business entity. The disadvantage is the higher start-up cost and the difficulty of raising capital. One form of business that may offer the advantages of both partnerships and corporations is the limited liability company (LLC). This form of business has a lower start-up cost, offers limited liability, and can be operated by partners or a single owner. See Figure 11-1 for the different forms of business.

Figure 11-1

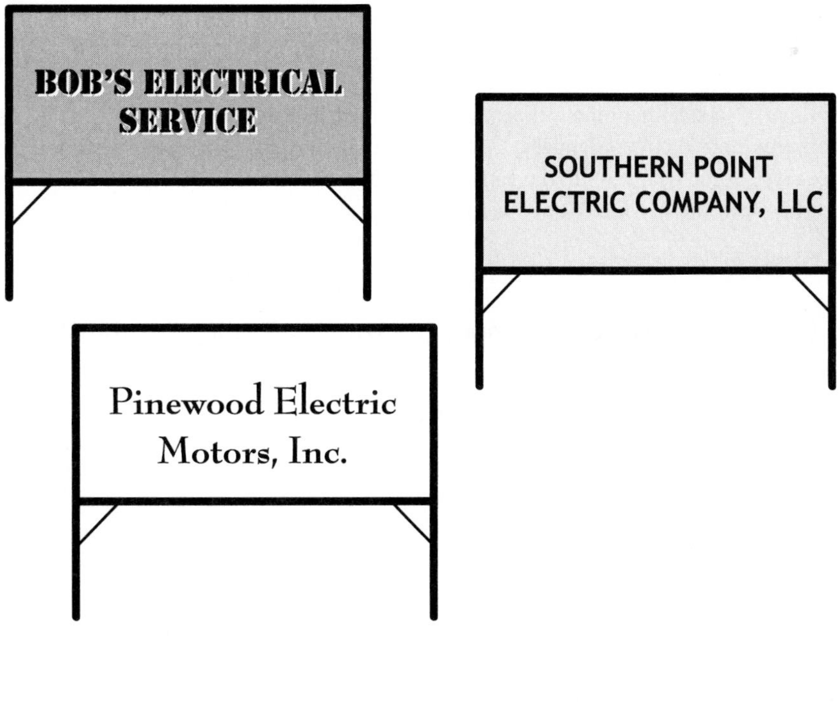

Types of Financial Statements

Income statement: a financial statement that summarizes and compares a company's **revenues** and expenses for a certain period. Refer to Table 11-1. The only items that should appear on the income statement should be revenues and expenses. Notice that the **net income** is the total revenue minus the total expenses.

Revenue — Income earned from the operations of a company

Net income — The result when revenue earned during an accounting period exceeds the expenses of the same period

Table 11-1: ABC Electrical Service Income Statement For the Month Ended June 30, 2005

Revenue:		
Service call revenue		$85,462.00
Contract revenue		$10,587.00
Expenses:		
Salary expense	$62,493.00	
Insurance expense	$5,000.00	
Tools and equipment expense	$2,185.00	
Vehicle expense	$3,850.00	
Total expense		$73,528.00
Net income:		$22,521.00

Statement of owner's equity: a statement that reflects increases or decreases in capital for a particular accounting period. In Table 11-2, the accounting period covers January 1, 2004, through January 31, 2004. If this company were to prepare a statement of **owner's equity** for the next month, the capital for February 1 would be $11,650.00. Notice that the increase in capital is the difference between the net income for the month and the withdrawals.

Owner's equity — Assets minus liabilities

Table 11-2: ABC Electrical Service Statement of Owner's Equity For the Month Ended January 31, 2004

Capital, Jan. 1, 2004		$9,400.00
Net income for the month	$6,500.00	
Less withdrawals	$4,250.00	
Increase in capital		$2,250.00
Capital, Jan. 31, 2004		$11,650.00

Net loss — The result when expenses exceed revenue during an accounting period

Balance sheet: a snapshot of company's assets, liabilities and owner's equity for a specific period of time. Assets (on the left) will equal liabilities and owner's equity (on the right). Notice that the total assets must equal the total liabilities plus the owner's equity. See Table 11-3.

Table 11-3: ABC Electrical Service Balance Sheet December 31, 2001

Assets		
Cash	$8,700.00	
Accounts Receivable	$1,500.00	
Inventory	$2,000.00	
Equipment	$17,200.00	
Supplies	$900.00	
Total Assets		$30,300.00

Liabilities		
Salaries payable		$6,400.00
Accounts payable		$7,250.00
Note payable		$2,000.00
Total Liabilities		$15,650.00

Owner's Equity		
Capital		$14,650.00
Total liabilities and owner's equity		$30,300.00

Cash — Assets such as paper money, coins, and checking accounts

Accounts receivable — Assets resulting from selling goods or services to customers with their promise to pay in the future

Equipment — Physical assets such as furniture, computers, and vehicles that are used in operation of the company

Accounts payable — Debt that results when goods or services are purchased on credit

Note payable — Written promissory note requiring the business to pay a certain amount in a specific amount of time (Example: mortgage payment)

You may or may not see an actual financial statement on an exam, but you may be asked to perform some calculations that will be contingent on your ability to understand and interpret basic financial statements. Later on in this chapter, you will have the opportunity to develop your skills in dealing with financial statement data.

Key Financial Ratios

What follows are some key financial ratios that may be encountered on an exam. These ratios are used to determine the financial strengths or weaknesses in a company. You can learn more about these basic ratios in *Basic Business and Project Management for Contractors*.

$$\text{Current ratio} = \frac{\textbf{Current Assets}}{\textbf{Current Liabilities}} \qquad \text{Quick ratio} = \frac{\text{Current Assets} - \text{Inventory}}{\text{Current Liabilities}}$$

Current asset — An asset that can be converted to cash within the period of a year (Example: cash, accounts receivable, prepaid rent, and inventory)

Current liability — A liability or debt that will be paid within the period of a year (Example: accounts payable, salaries payable, and federal taxes)

$$\text{Net worth} = \text{Assets} - \text{Liabilities}$$

$$\text{Percent Complete} = \frac{\text{Actual job cost}}{\text{Contract price estimate}}$$

$$\text{Working capital} = \text{Current Assets} - \text{Current Liabilities}$$

Important Things to Remember about Taxes

As a business owner, it is imperative to gain a comprehensive understanding of federal tax laws pertaining to employment. Failure to comply with federal tax laws could ultimately destroy a business. Much of this information can be collected from Circular E, which is produced by the federal government. The following are some important notes from Circular E referring to tax withholding. Note that the figures and the calculation examples used in this book are based upon 2006 tax information. The important thing to grasp from this chapter is how to properly use the information given to provide the correct solution. You should consult all the current year federal and state tax information that pertains to the exam you are taking.

- Tax withholding is calculated by using the Circular E tax table.
- Social Security is paid at a rate of 6.2%.
- Medicare is paid on all wages at a rate of 1.45%.
- The employee will have 6.2% deducted for Social Security and 1.45% deducted for Medicare. The employer has the responsibility to match the 6.2% Social Security as well as the 1.45% of Medicare. So, the total Social Security paid must be posted at 12.4% and Medicare at 2.9%.

Circular E, Withholding Practice Problems

Instructions: Obtain a copy of Circular E and calculate the following withholding questions.

Example 1

Example: An electrician's helper earns $7.00 an hour, is single, claims one withholding exemption, and is paid weekly. How much will he take home after a 40-hour work week?

    a. $251.00
    b. $239.58
    c. $280.00
    d. $221.36

$7.00 × 40 =	$280.00	Gross pay
$280.00 × 6.2% =	$ 17.86	Social Security tax
$280.00 × 1.45% =	$ 4.06	Medicare
Federal tax =	$ 19.00	Tax table in Circular E using the information provided: single, one exemption, and paid weekly
Take home	**$239.58**	

    1. A maintenance worker earns $15.75 an hour, is married, claims three exemptions, and is paid weekly. How much will be his take-home pay for a 40-hour week?
        a. $545.10
        b. $552.80
        c. $602.80
        d. $547.80

2. A counter salesman earns $8.00 an hour at Pinewood Electrical Supply. He is married, always works 45 hours a week, claims two exemptions, and is paid weekly. What will be his net pay?
   a. $340.93
   b. $315.48
   c. $296.89
   d. $321.45

3. An electrician is paid $18.00 per hour and works 40 hours in a week. She is single with no exemptions and is paid weekly. How much is her company required to pay in Medicare and Social Security?
   a. $45.64
   b. $10.44
   c. $64.58
   d. $55.08

More Circular E Practice Problems

Instructions: Obtain a copy of Circular E and answer the following withholding and labor law questions.

1. A Form 1099-Misc must be filed if you hire an independent contractor for extra help and pay wages that exceed _____ .
   a. $500
   b. $600
   c. $700
   d. $800

2. All employee records must be kept for at least _____ years.
   a. six
   b. three
   c. five
   d. four

3. A helper earns $8.00 per hour and works a total of 34 hours in a week. He is not married, claims no dependents, and is paid weekly. What is his take-home pay?
   a. $224.20
   b. $214.24
   c. $226.52
   d. $233.20

4. Supplemental wages include, but are not limited to, bonuses, commissions, overtime pay, payments for accumulated sick leave, severance pay, awards, prizes, _____ .
   a. gifts and rental reimbursement
   b. back pay and retroactive pay increases
   c. travel reimbursements
   d. mileage allowance and expenses

5. If a tax deposit was made 12 days late, the penalty rate would be _____ %.
   a. 10
   b. 5
   c. 2
   d. 15

6. Tommy is a professional ditch digger earning $9.00 an hour and is paid on a weekly basis. He is married, claims nine dependents, and always works 50 hours a week. What is his take-home pay?
   a. $456.13
   b. $445.52
   c. $457.13
   d. $495.00

Fair Labor Standards Act

The Fair Labor Standards Act provides federal guidelines to protect employees from unfair labor practices. It sets standards such as minimum wage, overtime compensation, and labor conditions. You should obtain a copy before completing the following questions.

1. Minimum wage is not less than _____ per hour, effective September 1, 1997.
   a. $5.15
   b. $5.00
   c. $4.75
   d. $5.25

2. The FLSA regulates the following practices:
   a. overtime pay, vacation, and sick pay
   b. pay raises and minimum wage
   c. number of employees and overtime pay
   d. minimum wage, overtime pay, and employment of minors

3. A construction worker works 48 hours one week. He earns $10.00 per hour. How much is his gross pay?
   a. $520.00
   b. $540.60
   c. $510.20
   d. $534.00

4. A tile installer has agreed to be paid $0.15 per tile and has 5,000 to install. It takes her a total of 50 hours to install the tile. What is her gross pay?
   a. $900.00
   b. $1,100.00
   c. $825.00
   d. $1,050.00

5. Youths 16 and 17 years old may perform _____ .
   a. any nonhazardous job for unlimited hours
   b. any nonhazardous job but limited to 40 hours a week
   c. any hazardous job but limited to 30 hours a week
   d. any type of work for unlimited hours

6. Under the FLSA, which of the following is not required to remain on file?
   a. total overtime pay
   b. number of vacation days available
   c. regular hourly pay rate
   d. birth date if under 19 years old

7. Violators of the minimum wage and overtime requirements can be fined up to _____ .
   a. $5,000
   b. $500
   c. $10,000
   d. $1,000

Business and Project Management

As mentioned earlier, some state exams use the *Basic Business and Project Management for Contractors* book on the exam while other states may use a specific version of *Business and Project Management for Contractors*. Regardless, the first nine chapters out of each state's book are typically the same and are relevant to all contractors. However, the chapters after nine usually cover information specific to each state. The questions used in this chapter cover information found in *Basic Business and Project Management for Contractors,* which is composed of nine chapters. Even though the state exam you are taking may not use *Basic Business and Project Management for Contractors* as an exam reference, it provides a good overview of basic business management and a good source of reference material.

1. The primary purpose of payment bonds is to _____ .
   a. protect the homeowner from damage resulting from subcontractors
   b. protect the property from liens resulting from unpaid subcontractors or suppliers
   c. guarantee that the contractor will complete the job
   d. guarantee that the contractor will complete the job according to the plans and specs.

2. Superintendents are responsible for all the following except _____ .
   a. day-to-day activities
   b. keeping job records
   c. filing claims for extra time
   d. project budgeting

3. One advantage of corporations is _____ .
   a. ease of establishment
   b. limited liability for debts and actions
   c. no income tax requirements
   d. the company can be liquidated quickly

4. Which of the following is not a primary element in making a contract?
   a. acceptance
   b. offer
   c. engagement
   d. consideration

5. The FLSA applies to companies or employers who employ _____ employees.
   a. one or more
   b. two or more
   c. 20
   d. 10

6. The financial ratio that reflects the relationship between debt and owner's equity is which of the following?
   a. Quick ratio
   b. Profit margin
   c. Activity ratio
   d. Debt/Equity ratio

7. When a subcontractor shares his or her bid amount with another subcontractor in order to achieve the lowest bid, it is known as _____ .
   a. bid rigging
   b. bid shopping
   c. builder's risk
   d. business trust

8. When referring to contract law, an offer usually consists of the following except _____ .
   a. scope of work
   b. negotiation
   c. price
   d. time schedule

9. The "Worker Adjustment and Retraining Notification Act" is applicable if a company employs _____ or more employees.
   a. 200
   b. 1,000
   c. 99
   d. 12

10. Employee personnel files are required to be maintained for at least _____ years.
   a. 4
   b. 7
   c. 10
   d. 2

Accounting Problems

There is no substitute for accurate and concise accounting methods. Everyone is not cut out to be a successful business analyst or accounting manager. Some are better suited for using their technical abilities while others make better business managers. Most exams don't go into a lot of accounting and bookkeeping scenarios, but if your state includes a business law section on the exam, you should have a basic understanding of accounting procedure. You can learn more about accounting in *Basic Business and Project Management for Contractors*.

Identifying Assets and Liabilities

The balance sheet in Table 11.4 demonstrates how assets = liabilities + owner's equity. Use the financial ratios described earlier to solve the equations.

Table 11-4: Pinewood Electrical Supply Co. Balance Sheet December 31, 2000

Assets	
Cash	$5,000
Accounts receivable	3,000
Supplies	1,500
Equipment	8,000
Total assets	**$17,500**

Liabilities	
Accounts payable	$2,000
Notes payable	2,500
Salaries payable	4,000
Total liabilities	**$8,500**

(continued on next page)

Table 11-4 *(continued)*

Owner's equity		
Capital	$9,000	
Assets − Liabilites =	**$17,500**	**Equity**

1. The current assets total to _____ .
   a. 17,500
   b. 8,000
   c. 9,500
   d. 5,000

2. The current liabilities total to _____ .
   a. 4,500
   b. 2,000
   c. 8,500
   d. 6,000

3. What is the current ratio?
   a. 1.58
   b. 1.34
   c. 0.58
   d. 1.86

4. What is the quick ratio?
   a. 1.58
   b. 1.33
   c. 1.20
   d. 2.05

5. Calculate the net worth of the company above.
   a. $9,000
   b. $17,500
   c. $26,000
   d. $8,000

Business Law Practice Exam

Use the three references listed to answer the following questions.

1. You have a helper who earns $6.50 per hour and works a total of 35 hours a week. What is his gross pay?
   a. $198.50
   b. $227.50
   c. $260.00
   d. none listed

2. Limited partnerships usually are what type of arrangement?
   a. single project
   b. 1 year term
   c. 3 to 5 projects
   d. none listed

3. By January 31st of each year, the employer must furnish which statements?
   a. W-4 and W-3
   b. Forms 8027 and 1096
   c. Forms 1099 and W-2
   d. Forms 1060 and 950

4. An electrician is paid $16.00 an hour and works 40 hours in a week. He is married with four exemptions and is paid weekly. How much is his company required to pay in Medicare and Social Security?
   a. $39.68
   b. $40.96
   c. $9.28
   d. $48.96

5. Form _____ is used to notify the IRS of address changes.
   a. 8822
   b. 8800
   c. 1040
   d. 1099

6. Employees under the age of _____ should not be employed in hazardous jobs.
   a. 16
   b. 18
   c. 15
   d. 21

7. Which method of accounting matches revenues to expenses and usually produces a more accurate depiction of the financial status of a company?
   a. cash basis
   b. completion basis
   c. accrual basis
   d. income basis

8. A payroll check is to be written for an employee who is paid $14.50 per hour and worked a total of 48 hours in a week. She is married, has two exemptions, and is paid every Friday. How much is her take-home pay?
   a. $754.00
   b. $639.32
   c. $715.43
   d. none listed

9. An employer is required to provide workers' compensation coverage if the company employs _____ employees.
   a. two or more
   b. 10
   c. 6
   d. three or more

10. An employee's wage base limit for 2006 is _____ .
    a. $94,200
    b. $80,000
    c. $100,000
    d. $87,000

Summary

Developing a thorough understanding of the many different aspects of business law would take years of education and experience. However, for most exams, establishing a basic knowledge of business law fundamentals is usually all that is required for passing the exam. Some exams may not contain any business law questions, but it is a good idea to gain as much knowledge as possible in order to increase the chances of becoming a successful electrical contractor. Be sure to become familiar with your state's business law requirements and use the most accurate information to solve any calculations. This chapter provides a general overview of how to perform some basic calculations. Take the time before the exam to review the current laws in your state as well as any federal wage and labor laws that you may encounter on the exam.

CHAPTER 12

THE PRACTICE COMPREHENSIVE EXAM

This chapter will allow you to demonstrate what you know and what you have learned from this text. The practice exam covers both the National Electrical Code as well as business law questions. There are 100 questions in this exam and you should be able to complete all of the questions within 6 to 8 hours. You should practice timing yourself to establish a good pace for the real exam. It will be very important on test day to be able to complete as many as possible if not all the questions in order to do well on the exam. As discussed in Chapter 1, try to answer as many questions as you can without looking through the Code. This will save you some valuable time. Also, remember to answer the easier questions first, leaving the more difficult questions for the end of the exam. You don't want to waste time trying to answer one question correctly while you leave 10 other easy questions unanswered because of a lack of time. Finally, remember you are shooting for at least a 70%. Good luck!

Directions

Use the answer sheet provided in the Appendix to record your responses. After you have completed the exam, use the detailed answer key in the Appendix to determine the number you have correct or locate the correct response. Please note that all the tax law questions are based upon 2006 information. You should note that tax regulations change each year so you should consult the most up to date tax information before taking the actual exam in your state. Using 2006 data is still effective in learning the proper techniques. Once you have mastered the techniques, you can simply plug in the current tax information.

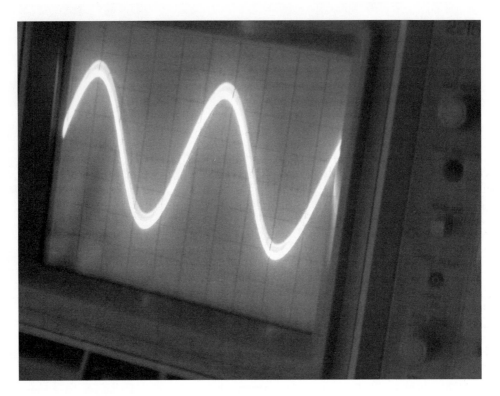

Practice Exam

1. The purpose of the NEC is to provide _____ .
   a. maximum protection
   b. practical safeguarding of persons
   c. instructions
   d. maximum standards

2. The NEC does not cover installations in _____ .
   a. carnivals
   b. floating buildings
   c. ships
   d. mobile homes

3. Who has the responsibility to interpret the Code?
   a. authority having jurisdiction
   b. state government
   c. home owner
   d. general contractor

4. A continuous load is a load in which the maximum current is expected to continue for _____ hours or more.
   a. 4
   b. 3
   c. 10
   d. 2

5. In an electrical control room, a condition exists where live exposed parts are present on one side and grounded parts on the other. The voltage is 480 V. How much working space is required?
   a. 4'
   b. 3'
   c. 3½'
   d. 2'

6. Receptacle outlets must be ground fault protected when located _____ from the outside edge of a wet bar sink.
   a. 6'
   b. 6½'
   c. 3'
   d. 8"

7. The rating of any one cord and plug connected utilization equipment shall not exceed _____ percent of the branch circuit amp rating.
   a. 70
   b. 85
   c. 80
   d. 5

8. Appliance outlets must be located within _____ feet of the intended location of the appliance.
   a. 6
   b. 10
   c. 3
   d. 8

9. The small-appliance circuits shall have no other outlets with the exception of a(n) _____ .
   a. smoke detector
   b. doorbell transformer
   c. electric clock
   d. outside light

10. A receptacle shall be located within _____ feet of HVAC equipment.
    a. 25
    b. 30
    c. 20
    d. 50

11. What is the resistance of 70 feet of #10 THHN solid conductor?
    a. 0.0847 ohms
    b. 0.537 ohms
    c. 1.24 ohms
    d. 0.742 ohms

12. Seven #12 THW conductors run through a conduit with an ambient temperature of 26 degrees Celsius. The ampacity of the conductor is _____ .
    a. 20 A
    b. 17.5 A
    c. 16 A
    d. 25 A

13. What is the ampacity of three #14 THWN conductors in a conduit with a ambient temperature of 40 degrees Celsius?
    a. 17 A
    b. 13.2 A
    c. 20 A
    d. 17.6 A

14. A square box contains two 12/2 cables and two 14/2 cables with grounds, one receptacle, and two clamps. One of the 12/2 cables in connected to the receptacle and the other cables only pass through the box. What is the cubic-inch requirement?
    a. 26
    b. 24.25
    c. 26.25
    d. 21.5

15. In straight pulls, the length of the box shall not be less than eight times the metric designator of the _____ raceway.
    a. smallest
    b. largest
    c. minimum
    d. maximum

16. Nonmetallic sheathed cable shall be supported at intervals not exceeding _____ inches of every box or cabinet.
    a. 24
    b. 8
    c. 6
    d. 12

17. Built-in dishwashers and trash compactors shall be permitted to be cord and plug connected; however, the _____ .
    a. receptacle does not have to be accessible
    b. length of the cord shall be 5 to 6 feet in length
    c. receptacle shall be located in the space occupied by the appliance
    d. cord is permitted to be directly wired

18. The branch circuit ampacity of a ½ hp, 230-volt, single-phase motor is _____ .
    a. 6.75 amps
    b. 5.4 amps
    c. 6.125 amps
    d. 4.9 amps

19. What size branch circuit overcurrent device is needed for a 2 hp, single-phase, 208-volt motor using a dual element fuse?
    a. 20 amp
    b. 25 amp
    c. 30 amp
    d. 35 amp

20. The disconnecting means for a motor shall be located _____ the driven machine.
    a. within 6' of
    b. within 20' of
    c. within sight of
    d. where accessible to

21. Class I locations are those locations that include _____ .
    a. flammable gases or vapors
    b. combustible dust and filings
    c. ignitable fibers
    d. explosive gases and mixtures

22. Mobile home parks are calculated at _____ volt amps per each mobile home lot.
    a. 1,500
    b. 20,000
    c. 12,000
    d. 16,000

23. What size circuit breaker is required for a 70-amp welder?
    a. 140 amp
    b. 150 amp
    c. 160 amp
    d. 70 amp

24. An arc welder has a duty cycle of 50% with a rated current of 60 amps. What is the required ampacity of the supply conductors?
    a. 42.6
    b. 45
    c. 60
    d. 120

25. The name plate on a resistance welder shall include the following except _____ .
    a. frequency
    b. primary voltage
    c. gap setting
    d. supply cord gauge

26. How much voltage is lost when using a #14 gauge, 150' drop cord for a load of 10 amps?
    a. 4.32 volts
    b. 7.6 volts
    c. 9.416 volts
    d. 4.35 volts

27. Accounts payable is an example of a(n) _____ .
    a. current asset
    b. current liability
    c. owner's equity
    d. revenue

28. What is the service demand for a 7 kW cooktop?
    a. 4.8 kW
    b. 6.0 kW
    c. 5.6 kW
    d. 7.0 kW

29. How much current does a 4-bulb, 120-volt fixture draw with each bulb rated at 60 watts?
    a. 2 amps
    b. 4 amps
    c. 2.5 amps
    d. 3 amps

30. What is the total rate for Social Security paid by the employee and employer?
    a. 6.2%
    b. 1.45%
    c. 2.9%
    d. 12.4%

31. If cables or wires are installed in notches in wood, the cable shall be protected by a _____ inch thick plate.
    a. $\frac{1}{16}$
    b. $\frac{1}{8}$
    c. $\frac{1}{4}$
    d. $\frac{1}{32}$

32. The location in which easily ignitable fibers are stored or handled other than in the process of manufacturing is known as a _____ .
    a. Class II, Div. I
    b. Class III, Div. 1
    c. Class III, Div. 3
    d. Class III, Div. 2

33. How many #12 conductors are allowed to be installed in a 4 $\frac{1}{16}$ × 2 $\frac{1}{8}$ square box?
    a. 12
    b. 18
    c. 10
    d. 16

34. What size "heater" is needed for a 2 hp, 208-volt, single-phase motor with a service factor of 1.15?
    a. 13.2 amps
    b. 18 amps
    c. 16.5 amps
    d. 12 amps

35. How many #12 THHN conductors are allowed in $\frac{3}{4}$" EMT?
    a. 12
    b. 14
    c. 18
    d. 16

36. An employee is allowed to be exposed to 100 decibels for a period of _____ hours per day.
    a. 8
    b. 2
    c. 4
    d. 3

37. Connections made between the grounding conductor and the grounding electrode may be all of the following except _____ .
    a. listed lugs
    b. listed clamps
    c. solder
    d. exothermic welding

38. How many amps would a neutral carry if L1 had a load of 16 amps and L2 had a load of 21 amps?
    a. 6
    b. 37
    c. 16
    d. 5

39. A series circuit contains two resistors. The first resistor has a resistance of 125 ohms and the second has a resistance of 75 ohms. What is the total resistance of the circuit?
    a. 200 ohms
    b. 50 ohms
    c. 125 ohms
    d. 100 ohms

40. Temporary electrical power shall be removed from a construction site _____ .
    a. within 90 days
    b. within 160 days
    c. after 10 days from the time of completion
    d. immediately after completion

41. A maintenance worker earns $12.00 per hour and works 40 hours per week. How much Social Security should be taken out of his pay?
    a. $34.52
    b. $29.76
    c. $57.60
    d. none listed

42. A minimum wage of not less than _____ an hour is allowed for employees under the age of 20 only for the first 90 consecutive days of employment.
    a. $5.25
    b. $4.75
    c. $4.25
    d. $4.50

43. A resistance welder has a duty cycle of 40% and a primary current of 40 amps. What is the amperage of the welder?
    a. 32.6 amps
    b. 25.2 amps
    c. 36.8 amps
    d. none listed

44. What is the lighting demand for a 10,000-square-foot bank according to 220.12?
    a. 30,000 VA
    b. 20,000 VA
    c. 40,000 VA
    d. 35,000 VA

45. General use dimmer switches can be used to control _____ .
    a. only permanently installed incandescent luminaires
    b. ceiling fans
    c. fractional horsepower motors
    d. any load unless otherwise stated

46. Busways must be marked with the following except for _____ .
    a. voltage rating
    b. current rating
    c. resistance rating
    d. manufacturer's name or trademark

47. A contractor has the following accounting entries:

    Cash $7,000
    Accounts Payable   2,000
    Equipment   4,000
    Notes Payable   2,000
    Accounts Receivable   3,000

    What are her total assets?
    a. $18,000
    b. $14,000
    c. $16,000
    d. $9,000

48. What is the ampacity of a #8 THW conductor installed in an attic with an ambient temperature of 130°F?
    a. 33.5 amps
    b. 50 amps
    c. 45 amps
    d. 37.5 amps

49. When calculating box fill, internal cable clamps, whether they are factory installed or installed by the installer, are _____ .
    a. counted as a double volume based on the largest conductor
    b. counted as a single volume based on the smallest conductor
    c. not counted in the calculation
    d. counted as a single volume based on the largest conductor

50. The maximum height that a switch or circuit breaker that is used as a switch can be located above the floor or working platform is _____ .
    a. 4'
    b. 5' 6"
    c. 6' 7"
    d. 7'

51. The four basic forms of business organization include all the following except _____ .
    a. partnerships
    b. corporations
    c. "T" corporations
    d. sole proprietorships
    e. "S" corporations

52. Which one of the following is not a common estimating method?
    a. cubic foot method
    b. conceptual method
    c. take-off method
    d. square foot method
    e. checklist method

53. What is the rated ampacity of a branch circuit conductor supplying a motor with an FLC of 17 amps?
    a. 21.25 amps
    b. 17 amps
    c. 24.5 amps
    d. 15.6 amps

54. Bonding conductors for electric signs shall be copper and not smaller than _____ AWG.
    a. 12
    b. 14
    c. 10
    d. 8

55. The area measuring 6 feet horizontally from a fuel dispensing pump is a _____ classified location.
    a. Class II, Div. 1
    b. Class I, Div. 1
    c. Class II, Div. 2
    d. Class I, Div. 2

56. What is the demand for a 14 kW electric range?
    a. 8 kW
    b. 10 kW
    c. 8.8 kW
    d. 11 kw

57. How many #12 THHN conductors can be installed in ¾" liquidtight flexible nonmetallic conduit?
    a. 16
    b. 22
    c. 13
    d. 10

58. What is the minimum size heater or overload device that should be installed on a 2 hp, 3-phase, 208-volt motor with a service factor of 1.2?
    a. 7.5 amps
    b. 9.375 amps
    c. 8.0 amps
    d. 12.5 amps

59. Edison-Base fuses shall be classified at not over _____ .
    a. 125 volts and 20 amps or less
    b. 125 volts and 40 amps or greater
    c. 240 volts and 30 amps or less
    d. 125 volts and 30 amps or less

60. Which of the following financial statements provides a "snapshot" of a company's assets, liabilities, and owner's equity?
    a. Income statement
    b. Statement of cash flows
    c. Balance sheet
    d. Working capital statement

61. What is the voltage drop on a 120-volt, single-phase circuit using a #12 copper conductor with a load of 7 amps and at a distance of 150' from the panel?
    a. 4.15 volts
    b. 3.62 volts
    c. 2.36 volts
    d. none listed

62. How many #8 AWG conductors are allowed to be installed in a 4" × 1½" square box?
    a. 10
    b. 8
    c. 6
    d. 7

63. Plaster, drywall, or plasterboard that is damaged around boxes that contain a flush type cover or a faceplate, shall be repaired so that no spaces are greater than _____ mm at the edge of the box.
    a. 6
    b. 5
    c. 4
    d. 3

64. An example of a long-term debt would include which of the following?
    a. mortgage
    b. income taxes
    c. unpaid wages
    d. accounts payable

65. A portable motor of _____ hp or less shall be permitted to utilize an attachment plug and receptacle.
    a. ½
    b. ¼
    c. ⅓
    d. 1

66. Which of the following parts of an atom contains a negative charge?
    a. Neutron
    b. Proton
    c. Electron
    d. Cruton

67. A series circuit contains two resistors. Resistor #1 has a value of 100 ohms and resistor #2 has a value of 150 ohms. What is the total resistance?
    a. 50 ohms
    b. 250 ohms
    c. 60 ohms
    d. 325 ohms

68. A 4/0 USE copper conductor has an ampacity of _____ amps.
    a. 200
    b. 195
    c. 260
    d. 230

69. How many 15-amp lighting circuits are required for a 2,100-square-foot home?
    a. 4 circuits
    b. 3 circuits
    c. 5 circuits
    d. 2 circuits

70. In a motor control circuit, what type of contact would be used for an emergency stop button?
    a. normally open
    b. normally closed
    c. momentary
    d. delay off

71. A salesman at Bob's Plumbing and Electrical Supply earns $9.00 per hour and works 40 hours each week and is single with no dependents. What is his take-home pay if he is paid weekly?
    a. $360.00
    b. $292.46
    c. $332.46
    d. $282.54

72. Rigid metal conduit shall be supported within _____ feet of each junction box, outlet box, device, or cabinet.
    a. 5
    b. 4
    c. 3
    d. 2

73. What is the demand for 12' of show window lighting that operates for 12 hours each day?
    a. 3,000 VA
    b. 2,400 VA
    c. 2,800 VA
    d. none listed

74. An office building contains a 20' section of multioutlet assembly which is used as general receptacles. What is the demand?
    a. 1,200 VA
    b. 3,600 VA
    c. 800 VA
    d. 720 VA

75. If the main bonding jumper is a screw only, then that screw shall be identified with a _____ .
    a. hex head
    b. GRD stamp
    c. green finish
    d. none listed

76. Identify which of the following pictures shows a pair of lineman pliers.

a.

b.

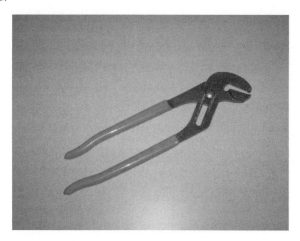

c.

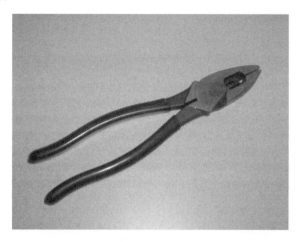

77. Which form of business has high personal liability but is the easiest to establish of the major forms of business?
   a. Partnership
   b. LLC
   c. Joint Venture
   d. Sole proprietorship

78. What type of insurance would protect against incidents such as fire, weather-related destruction, or theft?
   a. General liability
   b. Workers' Comp
   c. Property insurance
   d. Damage

79. When determining the general lighting load for a school, you should use _____ volts per square foot.
   a. 1
   b. 2
   c. 3
   d. 4

80. If seven pieces of commercial cooking equipment are to be installed, what is the allowable demand factor percentage?
   a. 80%
   b. 70%
   c. 90%
   d. 65%

81. Overhead spans of conductors such as services, shall not be less than _____ feet when located over a residential driveway.
   a. 12
   b. 15
   c. 10
   d. 18

82. Equipment grounding conductors that are smaller than a #6 AWG shall be _____ .
   a. type THW
   b. protected from physical damage by a raceway or cable armor
   c. solid type and insulated
   d. none listed

83. Which of the following is not a standard sized fuse or circuit breaker?
   a. 90 amp
   b. 350 amp
   c. 175 amp
   d. 55 amp

84. The Fair Labor Standards Act requires that _____ .
   a. the employer provide vacation pay
   b. the employer restrict maximum work hours to not more than 80 hours per week
   c. employers keep accurate time sheets
   d. employers provide lunch breaks

85. An emergency stop button on a piece of equipment should be a _____ push button.
   a. normally closed, recessed head
   b. normally open, recessed head
   c. normally open, mushroom head
   d. normally closed, mushroom head

86. When installing a panelboard on a cement block wall, the most appropriate anchoring device to use would be _____ .
    a. lag bolt
    b. toggle bolt
    c. concrete nails
    d. lead anchor

87. What article states that no gap or spaces in drywall shall be greater than ⅛" at the edge of the cabinet or box?
    a. 312.8
    b. 310.11
    c. 312.4
    d. 210.15

88. A security deposit or prepaid expense should be classified as a(n) _____ .
    a. long-term liability
    b. current liability
    c. other asset
    d. property and equipment

89. What is the branch circuit overcurrent size for a ¾ hp, 240-volt, single-phase motor using a dual element fuse?
    a. 7.6 amps
    b. 15 amps
    c. 13.3 amps
    d. 20 amps

90. What hazardous location is identified by the presence of easily ignitable fibers or flyings?
    a. Class I
    b. Class II
    c. Class III
    d. Class IV

91. Type metal clad cable shall not be permitted in which of the following conditions?
    a. in hazardous locations
    b. outdoors
    c. encased in concrete
    d. in a raceway

92. How many #8 THHN conductors are allowed to be installed in a section of ¾" electrical metallic tubing?
    a. 6
    b. 4
    c. 3
    d. 0

93. Optical fiber cables that are installed in hazardous locations shall comply with section _____ of the NEC.
    a. 250.12
    b. 250.78
    c. Table 770.154
    d. none listed

94. When installing underground cables near swimming pools, it is required that the underground cables be located no less than _____ feet from the inside wall of the pool unless it is necessary for the operation of the pool itself.
    a. 6
    b. 5
    c. 4
    d. 8

95. A nonmotor generator arc welder is to be installed with a duty cycle rating of 80% and a primary current rating of 60 amps. What is the required ampacity of the supply conductors?
    a. 53.4 amps
    b. 60 amps
    c. 57 amps
    d. 54.6 amps

96. Which type of accounting method recognizes income when it is earned before it is received and recognizes expenses when they are incurred whether they are actually paid or not?
    a. Cash basis
    b. Accrual basis
    c. Balance basis
    d. Income method

97. A(n) _____ load is when the wave form of the steady state current does not match the wave form of the applied voltage.
    a. nonlinear
    b. linear
    c. isolated
    d. volatile

98. Arc fault type circuit breakers are required to be installed in which of the following areas?
    a. Bathrooms and kitchens
    b. Kitchens only
    c. Bedrooms only
    d. Bedrooms and Baths

99. What is the total VA for 25 feet of track lighting?
    a. 1,750 VA
    b. 1,500 VA
    c. 1,475 VA
    d. 1,875 VA

100. When considering box fill, how are equipment grounding conductors calculated into the equation?
    a. single volume based on the smallest conductor in the box
    b. single volume based on the largest conductor in the box
    c. double volume based on the largest conductor in the box
    d. each grounding conductor is counted individually

CHAPTER ANSWER KEY

Chapter 1 Answers

Math Review Problems:

1. **300 mA**, First, convert 3/10 to a decimal = 0.3 Then, $0.3 \times 0.001 = $ **300 mA**

2. **285 micro amps**, 0.000285/0.000001 = **285 micro amps**

3. **0.045 kilo ohms**, 45/1000 = **0.045 kilo ohms**

4. **670,000 ohms**, $670 \times 1000 = $ **670,000 ohms**

5. **0.23 volts**, $230 \times 0.001 = $ **0.23 volts**

6. **0.00278 amps**, $2.78 \times 0.001 = $ **0.00278 amps**

More Math Practice Problems:

1. **1,060**

2. **3,760**

3. **103,770**

4. **42**

5. **2,125**

6. **24,000**

7. **B**, $12 \times 10 = $ **120**

8. **D**, $24 \times 12 \times 6 = $ **1,728**

9. **C**, $70 \times 45 = $ **3,150**

10. **D**, $2 \times 2 \times 1.5 = $ **6**

General Installation Practice Questions:

1. **A**, 210.11(C)(1)

2. **C, 6'6"**   210.52 (E)

3. **B, 600**   220.14(E)

4. **A, 230.10**

5. **D, 75**   240.6(A)

6. **A, 2**   250.52(A)(4)

7. **C, embedded in concrete**   334.12(A)(9)

8. **D, 14**   Table 344.30(B)(2)

Chapter 2 Answers

Ohm's Law Practice Problems:

1. **A, 1.33**   E/R= $120/90 = \mathbf{1.33}$

2. **C, 171**   E/I= $120/0.7 = \mathbf{171}$

3. **C, 500**   I × R= $.05 \times 10,000 = \mathbf{500}$

4. **A, 2.08**   P/E= $250/120 = \mathbf{2.08}$

5. **C, 12.5**   P/E= $3,000/240 = \mathbf{12.5}$

6. **B, 540**   E × I= $120 \times 4.5 = \mathbf{540}$

7. **B, 9 amps** $3 + 2.5 + 3.5 = \mathbf{9}$

8. **A, 13.33 ohms**   E/I = $120/9 = \mathbf{13.33}$ **ohms**

9. **C, 1.44 amps**   W/E = $400/277 = \mathbf{1.44}$ **amps**

10. **A, 133.3 mA**   E/R = $12/90 = .133$ amps converts to **133.3 mA**

Series Practice Problems:

1. **B, 2**   Since current remains the same in a series circuit, IT will be the same as I2 or **2 amps**.

2. **B, 250** $E = I \times R$ or $125 \times 2 = \mathbf{250}$

3. **D, 100**   First, add R1 and R2 together. Then subtract that from RT.
$75 + 125 = 200 \quad 200 - 300 = \mathbf{100}$

4. **A, 600** $E = I \times R$ or $2 \times 300 = \mathbf{600}$

5. **D**

Parallel Practice Problems:

1. **B, 85.71** $\dfrac{75 \times 125}{75 + 125} = \dfrac{9375}{200} = 46.875$

then,

$\dfrac{150 \times 200}{150 + 200} = \dfrac{30,000}{350} = \mathbf{85.71}$

2. **A, 90**   Voltage remains the same is a parallel circuit.

3. **D, 30**   RT = R/N $90/3 = \mathbf{30}$

4. **A, 45**   Voltage remains the same in a parallel circuit.

5. **B, 1.5**   I = E/R $45/30 = \mathbf{1.5}$

Voltage Drop Practice Problems:

1. **D, 2.24**   Use Table 8 for cm.

$\dfrac{2 \times 12.9 \times 90 \times 10}{10,380} = \dfrac{23,220}{10,380} = \mathbf{2.24 \ V}$

2. **B, #6** $\dfrac{2 \times K \times D \times I}{VD\ permitted} = \dfrac{2 \times 12.9 \times 185 \times 15}{3.6} = \dfrac{71,595}{3.6} = \mathbf{19,887.5}$

   Use Table 8 to find nearest conductor to 19,887.5
   #6

3. **A, 2.76 V** $\dfrac{2 \times 12.9 \times 100 \times 7}{6530} = \dfrac{18,060}{6530} = \mathbf{2.76\ V}$

4. **C, 1.93**   Table 8, page 636

5. **A, 7.457 amps** $\dfrac{2 \times 12.9 \times 250 \times 12}{10380} = \dfrac{77,400}{10,380} = \mathbf{7.457\ amps}$

6. **B, 7.2 V, yes** $240 \times 3\% = 7.2\%$ This voltage drop does exceed 7.2 V. The circuit is has a voltage drop of 7.457 volts which is greater than 7.2 volts allowed.

Chapter 3 Answers

Ampacity Practice Problems:

1. **B, 20 A**   A #12 copper conductor has a maximum ampacity of **20** amps according to 240.4(D).
   Table 310.16
   $25\ A \times 1.05 = 26.25$

2. **D, 38.5 A** $55\ A \times 70\% = \mathbf{38.5\ A}.$ Table 310.16 and Table 310.15(B)(2)(a)

3. **A, 30 A**   No derating is necessary because the conduit does not exceed 24". Article 240.4(D) restricts a #10 copper conductor to 30 amps. 310.15(B)(2)(a) Exception 3

4. **A, 55 A**   The ampacity of the conductor cannot exceed the temperature to which it is connected. See 110.14(C)

5. **D, 32.5 A** $65\ A \times 50\% = \mathbf{32.5\ A}$ Table 310.16 and Table 310.15(B)(2)(a)

6. **B, 12.18 A** $20A \times 0.87 = 17.4\ A \times 70\% = \mathbf{12.18\ A}$ Table 310.16 and Table 310.15(B)(2)(a)

7. **A, 10.92 A** $15\ A \times 0.91 = 13.65\ A \times 80\% = \mathbf{10.92\ A}$ Table 310.16 and Table 310.15(B)(2)(a)

8. **B, 30 A**   Table 310.16 See also 240.4 (D)

9. **B, 154.1 A**   A 4/0 USE conductor can carry 230 amps according to Table 310.16; however, it must be derated by 0.67 due to the ambient temperature. $230 \times 0.67 = \mathbf{154.1\ A}$

10. **B, 70**   Use Table 310.15(B)(2)(a) to locate the answer. Eight conductors in a raceway must be adjusted by **70%**.

Branch Circuit Loading and Conductor Ampacity Practice Problems:

1. **C, 3** $\dfrac{3\ VA \times 1,800\ sq\ ft}{15\ A \times 120\ V} = \dfrac{5,400}{1,800} = \mathbf{3\ circuits}$
   Table 220.3(A)

2. **C, 6** $\dfrac{180\ VA \times 80}{20\ A \times 120\ V} = \dfrac{14,400}{2,400} = \mathbf{6\ circuits}$
   220.3(B)(9)

3. **A, Yes**   **Yes** because the freezer does not exceed 80% of the circuit. 210.23(A)(1)

4. **D, 56.25 A** $25 \times 1.8\ A = 45\ A \times 125\% = 56.25\ A$ 210.19(A)(1)

5. **C, 3** $1.8\ A \times 120\ V = 216$ watts each   25 lights $\times$ 216 W = 5,400 watts
            $20\ A \times 120\ V \times 80\% = 1,920$ $\dfrac{5400}{1920} = 2.8$ or **3 circuits**

6. **D, 13** $\dfrac{180\ VA}{120\ V} = 1.5$ amps each

            $\dfrac{20\ A}{1.5} = 13.3$ or **13 receptacles**   220.3(B)(9)

7. **B, Yes, 80%**   210:23(A)(1)

8. **B, nonlighting outlet loads**   210.23(D)

9. **A, 12 amps**   Table 210.21(B)(2)

10. **C, not less than**   210.19(A)(2)

Chapter 4 Answers

Conduit Fill Practice Problems:

1. **B, ¾"**   Table C.1   p. 659

2. **D, 1¼"**   Table C.1   p. 669

3. **B, ¾"**   Table 5   p. 632
            #12 THHN = .0133 × 10 = .133
            #10 THHN = .0211 × 4 = .0844
                        .2174 needed

            Table 4   p. 627
                  Over 2 wires column
                  ¾"

4. **A, 56**   Table 4   p. 628
            Ex. 310.15 (B)(2)(A)
            1½"" RMC, 60% column = 1.188

            Table 5   p. 632
            #10 THHN = .0211 each

            $\dfrac{1.188}{.0211} = $ **56.30 conductors**

5. **C, 11**   Table C.9   p. 698

6. **C, 40%**   Table 1   p. 625

7. **C, 9**   Table C.5   p. 678

8. **A, 0.304 sq in**   Table 4   p. 626

9. **D, 60**   Table 1, Note 4   p. 625

10. **B, 0.8**   Table 1, Note 7   p. 625

Box Fill Practice Problems:

1. **C, 5**　　　　　Table 314.16(A)

2. **D, 21.5**　　　Table 314.16(A)

3. **A, (3" x 2" x 3½")**　Table 314.16(A)

4. **D, 17**　　　　Table 314.16(B)
#14	$2.00 \times 4 = 8.00$
#12	$2.25 \times 2 = 4.50$
Grounds	$2.25 \times 1 = 2.25$
Clamps	$2.25 \times 1 = \underline{2.25}$
	17

5. **D, 23.75**　　　Table 314.16(B)
#14	$2.00 \times 2 = 4.00$
#12	$2.25 \times 2 = 4.50$
Rec.	$2.25 \times 2 = 4.50$
Switch	$2.00 \times 2 = 4.00$
Grounds	$2.25 \times 1 = 2.25$
Clamps	$2.25 \times 1 = 2.25$
Stud	$2.25 \times 1 = \underline{2.25}$
	23.75

6. **A, 1**　　　　　314.16 (B) (1)

7. **C, double volume based on the largest conductor connected to it**　　　314.16 (B) (4)

Chapter 5 Answers

Lighting Demand Practice Problems:

1. **C, 5,310 VA**

sq ft	$1,700 \times 3$ VA	=	5,100 VA	Table 220.12
Sm. App.	$1,500 \times 2$	=	3,000 VA	210.11 (C) (1)
Laundry	$1,500 \times 1$	=	1,500 VA	210.11 (C) (2)
			9,600 VA	

Table 220.42		
First 3,000 VA @ 100%	=	3,000 VA
Remainder 6,600 VA @ 35% =		2,310 VA
		5,310 VA

2. **A, 4,995 VA**

sq ft	$1,400 \times 3$ VA	=	4,200 VA	Table 220.12
Sm. App.	$1,500 \times 2$	=	3,000 VA	210.11 (C) (1)
Laundry	$1,500 \times 1$	=	1,500 VA	210.11 (C) (2)
			8,700 VA	

Table 220.42		
First 3,000 VA @ 100%	=	3,000 VA
Remainder 5,700 VA @ 35% =		1,995 VA
		4,995 VA

3. **B, 4,838**

1,850 – 600 unfinished basement = 1,250 sq ft				
sq ft	1,250 × 3 VA	=	3,750 VA	Table 220.12
Sm. App.	1,500 × 2	=	3,000 VA	210.11 (C) (1)
Laundry	1,500 × 1	=	1,500 VA	210.11 (C) (2)
			8,250 VA	

Table 220.42
First 3,000 VA @ 100%	=	3,000 VA
Remainder 5,250 VA @ 35% =		1,838 VA
		4,838 VA

Cooktops, Ovens, and Ranges Practice Problems:

1. **B, 7.35 kW**

   $3.5 \times 3 = 10.5$ @ 70% = **7.35 kW**

2. **A, 8.4 kW**

   Use Column C
   8 + 5% = **8.4 kW**

3. **C, 17.2 kW**

   $6 \times 8 = 40$ kW @ 43% = **17.2 kW**

4. **B, 8.4 kW**

3.5 kW @ 80% =	2.8 kW	Column A
7 kW @ 80% =	5.6 kW	Column B
	8.4 kW	

5. **B, 8.8 kW**

   4 kW + 4 kW + 6 kW = 14 kW
   Use Column C   Since 14 kW is greater than 12 kW by 2 kW, add 10%
   8 kW + 10% = **8.8 kW**

6. **D, 11 kW**

2×3 kW = 6 kW @ 75% =	4.5 kW	Column A
2×5 kW = 10 kW @ 65% =	6.5 kW	Column B
	11 kW	

Dryers Practice Problems:

1. **B, 25,500 W**

   6,000 W × 5 = 30,000 w @ 85% = **25,500 W**

2. **C, 20,000 W**

   Use 5,000 W instead of 3,500 W. See 220.54
   5,000 W × 4 = 20,000 W @ 100% = **20,000 W**

3. **A, 32,500 W**

   6,500 W × 10 = 65,500 W @ 50% = **32,500 W**

4. **D, 38 kW**

   = 47 – (20–11)
   = 47 – (9)
   = 38%
   20 × 5 kW = 100 kW
   100 kW @ 38% = **38 kW**

5. **B, 55 kW**

   44 × 5 kW = 220 kW @ 25% = **55 kW**

Appliance Load Practice Problems:

1. **B, 9,960 W**

4,000 W dishwasher	
960 W pool pump	$(240 \times 4 = 960 \text{ W})$
5,000 W water heater	
9,960 W	Do not count the dryer; it has to be calculated separately using Table 220.54

2. **A, 8,145 W**

300 W disposal	$(120 \times 2.5 = 300 \text{ W})$
5,000 W water heater	
360 W compactor	$(120 \times 3 = 360 \text{ W})$
1,200 W pool pump	$(240 \times 5 = 1,200 \text{ W})$
4,000 W dishwasher	
10,860 @ 75% = **8,145 W**	

Grounding Electrode Conductor Practice Problems:

1. **B, #6**

2. **B, the equivalent size of the largest service entrance conductor required for the served load**

3. **B, #4**

Chapter 6 Answers

Commercial Lighting Load Practice Problems:

1. **B, 43,750 VA**

Table 220.42
10,000 VA × 3.5 = 35,000 VA @ 100% = 35,000 VA × 125% = **43,750 VA**

2. **A, 24,063 VA**

Table 220.42
5,500 VA × 3.5 = 19,250 VA @ 100% = 19,250 VA × 125%
= **24,063 VA**

3. **D, 15,000 VA**

Table 220.42
12,000 VA × 1 = 12,000 VA @ 100% = 12,000 VA × 125%
= **15,000 VA (not listed)**

4. **C, 11, 125 VA**

Table 220.42	
10,000 VA × 2 =	20,000 VA
2,000 VA × ½ = 1,000 VA × 125% =	1,125 VA
	21,125 VA
First 20,000 VA @ 50% =	10,000 VA
Hallway	1,125 VA
	11,125 VA

Notice how only the 20,000 VA was derated with the first 20,000 VA @ 50%. Hallways, which are considered "all others", are calculated at 100% according to Table 220.42 so therefore only the guest rooms are derated. You add the hallway back to the derated guest room calculation to get the total.

5. **A, 46,063 VA** $15 \times 25 = 375 \times 125 = 46,875$ sq. ft. (rooms)
$75 \times 10 = 750 \times 7 = 5,250$ sq. ft. (hallway)

Table 220.42
$46,875 \times 2 =$	93,750 VA
$5,250 \times 1 = 5,250 \times 125\% =$	6,563 VA
	100,313 VA

First 20,000 VA @ 50% =	10,000 VA
Remainder 73,750 VA @ 40%=	29,500 VA
	39,500 VA
Hallway	+6,563 VA
	46,063

Notice how only the 93,750 VA was derated with the first 20,000 @ 50%. The remainder of the 93,750, which was 73,750VA, was then derated by 40%. Hallways, which are considered "all others", are calculated at 100% according to Table 220.42 so therefore only the guest rooms are derated. You add the hallway back into the derated guest room calculation to get the total.

Receptacles, Multioutlet Assemblies, and Show Window Lighting Problems:

1. **B, 10.85 kVA**   220.14 (L)
65×180 VA = 11,700 VA or 11.7 kVA
Table 220.44
First 10 kVA @ 100%	=	10 kVA
Remainder of 1.7 kVA @ 50% =		.85 kVA
		10.85 kVA

2. **D, 18.05 kVA**   220.14 (L)
145×180 VA = 26,100 VA or 26.1 kVA
Table 220.44
First 10 kVA @ 100%	=	10 kVA
Remainder of 16.1 kVA @ 50% =		8.05 kVA
		18.05 kVA

3. **B, 3.6 amps**   220.14(H)
$12/5 = 2.4 \times 180$ VA = 432/120 V = **3.6 A**

4. **A, 9 amps**   220.14(H)
$10/5 = 2 \times 180$ VA = 360/120 V = **9 A**

5. **C, 7,500 VA**   220.14(G)
$30 \times 200 = 6,000 \times 125\% =$ **7,500 VA**

Commercial Cooking Equipment Practice Problems:

1. **A, 14.85 kVA**   220.56 and Table 220.56
6 kVA + 5 kVA + 5.5 kVA = 16.5 kVA × 90% = **14.85 kVA**

2. **C, 23.08 kVA**   5.5 kVA + 6 kVA + 6 kVA + 4.5 kVA + 4.5 kVA + 4 kVA
+ 5 kVA = 35.5 kVA × 655 = **23.08 kVA**

3. **B, 90%**   Table 220.56   Note that the gas stove should not be included in the calculation, so only three pieces of equipment should be counted.

4. **B, 18 kW**   Table 220.56   If two pieces of equipment are being considered, the demand should be calculated at 100%.

5. **A, less than, largest**   220.56

6. **B, 24.4 kW**   10 kW + 8 kW + 5 kW + 7.5 kW = 30.5 kW × 80% = 24.4 kW

Chapter 7 Answers

Motor Practice Problems:

1. **B, 12 A**   Table 430.248

2. **A, 12 A**   Table 430.248 and 430.32(A)(1)
   9.6 FLC × 125% = **12 amps**

3. **D, #8**   Table 430.248 and 430.22(A)
   28 FLC × 125% = 25 amps
   Refer to Table 310.16 under THHN column for wire size

4. **C, 30 A**   Table 430.248 and Table 430.52
   17 FLC × 175% = 29.75 amps
   Choose next higher size from 240.6(A)

5. **A, 35 A**   Table 430.248 for FLC of both motors and 430.24
   12 FLC for 2 hp motor
   17 FLC for 3 hp motor
   (17 × 125%) + 12 =   21.25 + 12 = 33.25 amps
   Choose next higher size from 240.6(A)

6. **C, 45 A**   430.62(A)
   25 A + 10 + 10 = **45 amps**

Three-Phase Motor Practice Problems:

1. **B, 110 A**   430.52 and Table 430.250
   59.4 FLC × 175% = 103.95 amps
   Choose next higher size from 240.6(A)

2. **D, #8**   430.22 and Table 430.250
   34 FLC × 125% = 42.5 amps
   Refer to Table 310.16 under 75 degree column for wire size

3. **B, #10**   430.24 and Table 430.250
   Single phase motor = 8 FLC
   Three phase motor = 10.6 FLC
   (10.6 × 125%) + 8 =   13.25 + 8 = 21.25 amps
   Refer to Table 310.16 under 60 degree column for wire size

4. **B, 60 A**   430.62(A) and Table 430.250
   16.7 FLC × 175% = 29.225 amps or 30 amp fuse
   30 + 16.7 + 16.7 = 63.4 amps, round down to next lower size

5. **A, 30.25 A**   430.32 and Table 430.250
   24.2 FLC × 125% = **30.25 amps**

Chapter 8 Answers

Transformer Calculation Practice Problems:

1. **B, 20 A**   Ratio = 1:2 so 40/2 = **20 amps**

2. **C, 7 A**   25 – 18 = **7 amps**

3. **C, 120 V**   Ratio = 4:1 because 40/10 = 4 Therefore, 480/4 = **120 volts**

4. **A, 4,800 VA**   Input = Output

Three-Phase Practice Problems:

1. **D, 20 A**   I line = I phase

2. **A, 360 V**   E line = E phase × 1.732

3. **C, 34.6 A**   I phase = I line/1.732

4. **B, 83,136 VA**   1.732 × E line × I line

                       1.732 × 240 × 200 = **83,136 VA**

5. **A, 240 V**   E line = E phase

6. **D, 21,840 VA**   3 × E phase × I phase

                       3 × 208 × 35 = **21,840 VA**

Chapter 9 Answers

Motor Control Practice Problems:

1. **D, both B and C**

2. **B, limit**

3. **D, foot**

4. **A, emergency stop**

5. **B, M1**

6. **A, NO**

7. **B, PB5**

8. **B, NC**

9. **A, PB5**

10. **C, pressing the start on line 5**

11. **D, M2, parallel**

Chapter 10 Answers

Mobile Home Practice Problems:

1. **D, 23**   Table 550.31

2. **A, 260 A** $16,000 \times 15 \times 26\%/240 =$ **260 A**   550.31 and Table 550.31

3. **C, 1,650 A** $20,000 \times 90 \times 22\%/240 =$ **1,650 A**   550.31 and Table 550.31

4. **D, rigid or IMT conduit**   550.15(H)

5. **C, 100**   550.32(C)

Welder Practice Problems:

1. **A, 110** $65 \times 0.84 = 54.6$ amps $\times 200\% = 109.2$ amps   Table 630.11(A)
   Refer to 240.6 for breaker size. The next higher breaker is **110 amp**.

2. **C, 28.4 A** $40 \times 0.71 =$ **28.4 amps**   Table 630.31(A)(2)

3. **B, 125 A** $40 \times 300\% = 120$ amps   630.32(A) and 240.6 Since there is no 120 amp breaker, you
   must choose the next higher size of **125 amps**.

4. **D, 159.75 A**   Table 630.11(A) and 630.11(B)
   $60 \text{ A} \times 0.75 \times 100\% = 45$
   $60 \text{ A} \times 0.75 \times 100\% = 45$
   $60 \text{ A} \times 0.75 \times 85\% = 38.25$
   $60 \text{ A} \times 0.75 \times 70\% = 31.5$
   159.75 amps

5. **A, 66.75 A** $75 \times 0.89 =$ **66.75 amps**   Table 630.11(A)

6. **B, 139.16 A** $70 \text{ A} \times 0.71 = 49.7$ Table 630.11(A)
   $60 \text{ A} \times 0.71 \times 60\% = 25.56$
   $50 \text{ A} \times 0.71 \times 60\% = 21.3$
   $50 \text{ A} \times 0.71 \times 60\% = 21.3$
   $50 \text{ A} \times 0.71 \times 60\% = 21.3$
   139.16 amps

Chapter 11 Answers

Circular E, Withholding Practice Problems:

1. **B, $552.80**

15.75×40	=	$630.00
$-630.00 \times 6.2\%$	=	39.06
$-630.00 \times 1.45\%$	=	9.14
$-$ Tax	=	29.00
		$552.80

2. **A, $340.93**

$$
\begin{aligned}
8.00 \times 40 &= \$320.00 \\
+OT\ 12.00 \times 5 &= 60.00 \\
-380.00 \times 6.2\% &= 23.56 \\
-380.00 \times 1.45\% &= 5.51 \\
-\text{Tax} &= 10.00 \\
\hline
&\ \$340.93
\end{aligned}
$$

3. **D, $55.08**

$$
\begin{aligned}
18.00 \times 40 &= \$720.00 \\[4pt]
720.00 \times 6.2\% &= 44.64 \\
+720.00 \times 1.45\% &= 10.44 \\
\hline
&\ \$\,55.08
\end{aligned}
$$

More Circular E Practice Problems:

1. **B, $600**

2. **D, four**

3. **A, $224.20**

$$
\begin{aligned}
8.00 \times 34 &= \$272.00 \\
-272.00 \times 6.2\% &= 16.86 \\
-272.00 \times 1.45\% &= 3.94 \\
-\text{Tax} &= 27.00 \\
\text{Take home pay} &= \$224.20
\end{aligned}
$$

4. **B, back pay and retroactive pay increases**

5. **B, 5**

6. **C, $457.13**

$$
\begin{aligned}
9.00 \times 40 &= 360.00 \\
OT\ 13.5 \times 10 &= 135.00 \\
\text{Total} &= \$495.00 \\
-495.00 \times 6.2\% &= 30.69 \\
-495.00 \times 1/45\% &= 7.18 \\
-\text{Tax} &= 0 \\
\text{Take home pay} &= \$457.13
\end{aligned}
$$

Fair Labor Standards Act:

1. **A, $5.15**

2. **D, minimum wage, overtime pay, and employment of minors**

3. **A, $520.00**

$$
\begin{aligned}
10.00 \times 40 &= \$400.00 \\
OT\ 15.00 \times 8 &= 120.00 \\
\text{Gross pay} &= \$520.00
\end{aligned}
$$

4. **C, $825.00**

$0.15 \times 5000 = 750.00$ total regular pay
To calculate overtime pay, you take the total pay and divide by the total hours
$750.00/50 = 15.00$ per hour
Take ½ of the regular rate of 15.00, which is 7.50 and multiply by the number of overtime hours
$7.50 \times 10 = 75.00$
$750.00 + 75.00 = \$825.00$

5. **A, any nonhazardous job for unlimited hours**

6. **B, Social Security number**

7. **D, $1,000**

Business Project Management

1. **B**   2-7
2. **D**   3-1
3. **B**   1-4
4. **C**   9-3
5. **A**   7-1
6. **D**   8-18
7. **A**   3-7
8. **B**   9-3
9. **C**   7-10
10. **A**   7-11

Accounting Problems

1. **C**   Cash + A/R + Supplies
2. **D**   A/P + Salaries payable
3. **A**   9,500/6,000 = 1.58
4. **B**   8,000/6,000 = 1.33
5. **A**   17,500 − 8,500 = 9,000

Business Law Practice Exam

1. **B**
2. **A**   1-5   BPMG
3. **C**   7-8   BPMG
4. **D**
5. **A**
6. **B**
7. **C**
8. **B**
9. **D**
10. **A, $94,200**   Page 16 Circular E

Chapter 12 Answers to Practice Exam

Answer Key:

FLSA = Fair Labor Standards Act
BBPMC = Basic Business and Project Management for Contractors

1. **B, practical safeguarding of persons**   90.1 (A)

2. **C, ships**   90.2 (B)(1)

3. **A, authority having jurisdiction** 90.4

4. **B, 3**   Article 100 p. 70

5. **C, 3 ½"**   Table 110.26 (A)(1) Condition 2

6. **A, 6'**   210.8 (A)(7)

7. **C, 80**   210.23 (A)(1)

8. **A, 6**   210.50 (C)

9. **C, electric clock**   210.52 (B)(2) Exception #1

10. **A, 25**   210.63

11. **A, 0.0847 ohms** $1.21/1000' = .00121 \times 70' = .0847$ Table 8

12. **D, 14 A**   Table 310.16 says #12 THW will carry 20 amps maximum
    Step 1: Refer to 240.4 (D)
    Step 2: Refer to correction factors on Table 310.16 for temperature
    Step 3: Table 310.15 (B)(2)(a) since there are more than 3 conductors
        $20 A \times 1.00 \times 70\% = 14$ amps

13. **B, 13.2 amps**
    Table 310.16 = 15 amps per 240.4 (D)
    $15 \times .88 = $ **13.2 amps**

14. **A, 26 in**
    Use Table 314.16 (B) to find in of each conductor. Refer to 314.16 (B)(1) & (5)
4 - #12	$2.25 \times 4 =$	9.00
4 - #14	$2.00 \times 4 =$	8.00
1 – rec.	$2.25 \times 2 =$	4.50
1 – clamp	$2.25 \times 1 =$	2.25
1 – ground	$2.25 \times 1 =$	2.25
		26 in

15. **B, largest**   314.28 (A)(1)

16. **D, 12**   334.30

17. **C, receptacle shall be located in the space occupied by the appliance**   422.16(B)(2)

18. **D, 4.9 amps**   Table 430.248

19. **B, 25 amps**   Table 430.248
    FLC = 13.2
    Table 430.52 under dual element
    $13.2 \times 175\% = 23.1$
    Refer to 240.6 (A) for next highest size which is **25 amps**

20. **C, within sight of**   430.102 (B)

21. **A, flammable gases or vapors**   500.5 (B)

22. **D, 16,000**   550.31

23. **B, 150 amps**   630.12 (A)
    70 A × 200% =140   Refer to 240.6 for next highest size which is 150 amps

24. **A, 42.6**   630.11 and Table 630.11 (A)
    60 A × 0.71 = **42.6 amps**

25. **D, supply cord gauge**   630.34

26. **C, 9.416 V**   Use Table 8 $\dfrac{2 \times 12.9 \times 150 \times 10}{4110} = \dfrac{38,700}{4110} = \textbf{9.416 V}$

27. **B, current liability**   BBPMC   8-9

28. **C, 5.6 kW**   80% × 7 kW = 5.6   Use Col. B in Table 220.55

29. **A, 2 amps**   4 × 60 W = 240 watts   I = P/V = 240/120 = 2 amps

30. **D, 12.4%**   Circular E

31. **A, 1/16"**   300.4(A)(2)

32. **D, Class III, Div. 2**   500.5(D)(2)

33. **B, 18**   Table 314.16(A)

34. **C, 16.5 amps**   Step 1 Table 430.248   FLC = 13.2 amps
    Step 2 Table 430.32(A)(1)   125%
    Step 3 13.2 × 125% = **16.5 amps**

35. **D, 16**   Table C1 on page 659

36. **B, 2**   BBPMC

37. **C, solder**   25.70

38. **D, 5**   21-16 = **5 amps**

39. **A, 200 ohms**   125 + 75 = **200 ohms**

40. **D, immediately after completion**   590.3

41. **B, $29.76**   12 × 40 = $480.00   480.00 × 6.2% = **$29.76**   FLSA

42. **C, $4.25**   FLSA

43. **B, 25.2 amps**   40A × 0.63 = **25.2A**   Table 630.31(A)(2)

44. **D, 35,000**   10,000 × 3½ = **35,000**   Table 220.12

45. **A, only to control permanently installed incandescent luminaries**   404.14(E)

46. **C, resistance rating**   368.120

47. **B, $14,000**   Cash + Equity + A/R

48. **A, 33.5 amps**   50 × 0.67 = 33.5   Table 310.16

49. **D, a single volume based on the largest conductor**   314.16(B)(2)

50. **C, 6'7"**   404.8(A)

51. **C, T corporations**   BBPMC   1-3

52. **E, checklist method**   BBPMC   4-4

53. **A, 21.25 amps**   17 × 125% = **21.25**   430.22

54. **B, 14**   600.7(D)

55. **D, Class I, Div. 2**   Fig. 514.3 NEC

56. **C, 8.8 kW**   8 kW + 10% = **8.8 kW**   Table 220.55 Col. C   Note 1

57. **A, 16**   Table C.5   p. 679

58. **B, 9.375 amps**   FLC = 7.5 7.5 × 125% = **9.375**   Table 430.250
See 430.32(1)

59. **D, 125 volts and 30 amps or less**   240.51(A)

60. **C, Balance sheet**   BBPMC   8-7

61. **A, 4.15 volts** $\dfrac{2 \times K \times D \times I}{\text{Cir. Mils}} = \dfrac{2 \times 12.9 \times 150 \times 7}{6530} = \dfrac{27,090}{6530} = \textbf{4.148 volts}$

Table 8 p.635

62. **D, 7**   Table 314.16(A)

63. **D, 3**   314.21

64. **A, mortgage**   BBPMC   8-8

65. **C, 1/3 hp**   430.81(B)

66. **C, Electron**

67. **B, 250 ohms**   Rt = R1 + R2 = 100 + 150 = **250 ohms**

68. **D, 230**   Table 310.16

69. **A, 4 circuits** $\dfrac{3 \text{ VA} \times 2100}{15 \times 120} = \dfrac{6300}{1800} = \textbf{3.5 or 4}$

Table 220.12

70. **B, normally closed**

71. **B, $292.46**
        9.00 × 40   =   360.00
        360 × 6.2% =   22.32
        60 × 1.45% =   5.22
        Tax   =   40.00
                $292.46

72. **C, 3**   344.30(A)

73. **A, 3,000 VA**   200 × 12' = 2,400 x 125% = **3,000 VA**   220.14(G)   215.2(A)(1)

74. **D, 720 VA**   20'/5' = 4 × 180 VA = **720 VA**   220.14(H)(1)

75. **C, green finish**   250.28(B)

76. **C**

77. **D, sole proprietorship**   BBPMC 1-3

78. **C, property insurance**   BBPMC 2-3

79. **C, 3**   Table 220.12   p. 58

80. **D, 65%**   Table 220.56   p. 61

81. **A, 12**   225.18(2)

82. **B, protected from physical damage by a raceway or cable armor**   250.122

83. **D, 55 amp**   240.6

84. **C, employers keep accurate time sheets**   BBPMC 7-2

85. **C, normally open, mushroom head**

86. **D, lead anchor**   *Rational:* All spaces in a block wall are not hollow for a toggle bolt to fasten.

87. **C, 312.4**

88. **C, other asset**   BBPMC   8-8

89. **B, 15 amps**   7.6 FLC × 175% = 13.3 amps   Choose next highest according to 240.6   Table 230.248

90. **C, Class III**   500.5(D)

91. **C, encased in concrete**   330.12

92. **A, 6**   Table C1   p. 659

93. **C, Table 770.154**

94. **B, 5**   680.10

95. **A, 53.4 amps**   60 A × 0.89 = 53.4 amps   Table 630.11(A)

96. **B, accrual basis** BBPMC   8-5

97. **A, nonlinear**   Article 100 p. 30

98. **C, Bedrooms only**   210.12

99. **D, 1875 VA**   **25/2 = 12.5 × 150 VA = 1875 VA**   **220.43(B)**

100. **B, single volume based on the largest conductor in the box**   314.16(B)(5)

B

PRACTICE EXAM
ANSWER SHEETS

ANSWER SHEET

Directions for Marking the Answer Sheet

- Use a No. 2 lead pencil. Do **NOT** use ink or ball point pen.
- Make dark marks that completely fill the circle.
- Make **NO** stray marks on the answer sheet.

A NAME (Please print clearly.)

Last Name First Name Middle Initial

B FIRST 4 LETTERS OF LAST NAME

Ⓐ Ⓑ Ⓒ Ⓓ Ⓔ Ⓕ Ⓖ Ⓗ Ⓘ Ⓙ Ⓚ Ⓛ Ⓜ Ⓝ Ⓞ Ⓟ Ⓠ Ⓡ Ⓢ Ⓣ Ⓤ Ⓥ Ⓦ Ⓧ Ⓨ Ⓩ

C APPLICANT ID NUMBER

⓪ ① ② ③ ④ ⑤ ⑥ ⑦ ⑧ ⑨

D BIRTHDAY

MONTH	DAY	
Jan.		
Feb.		
Mar.	⓪	⓪
Apr.	①	①
May	②	②
June	③	③
July		④
Aug.		⑤
Sept.		⑥
Oct.		⑦
Nov.		⑧
Dec.		⑨

E HAVE YOU EVER TAKEN THIS EXAM BEFORE?

Yes No

F TEST FORM NUMBER

⓪ ① ② ③ ④ ⑤ ⑥ ⑦ ⑧ ⑨

G TEST CENTER NUMBER

⓪ ① ② ③ ④ ⑤ ⑥ ⑦ ⑧ ⑨

TEST BOOKLET NUMBER

1 Ⓐ Ⓑ Ⓒ Ⓓ	41 Ⓐ Ⓑ Ⓒ Ⓓ	81 Ⓐ Ⓑ Ⓒ Ⓓ	121 Ⓐ Ⓑ Ⓒ Ⓓ	161 Ⓐ Ⓑ Ⓒ Ⓓ
2 Ⓐ Ⓑ Ⓒ Ⓓ	42 Ⓐ Ⓑ Ⓒ Ⓓ	82 Ⓐ Ⓑ Ⓒ Ⓓ	122 Ⓐ Ⓑ Ⓒ Ⓓ	162 Ⓐ Ⓑ Ⓒ Ⓓ
3 Ⓐ Ⓑ Ⓒ Ⓓ	43 Ⓐ Ⓑ Ⓒ Ⓓ	83 Ⓐ Ⓑ Ⓒ Ⓓ	123 Ⓐ Ⓑ Ⓒ Ⓓ	163 Ⓐ Ⓑ Ⓒ Ⓓ
4 Ⓐ Ⓑ Ⓒ Ⓓ	44 Ⓐ Ⓑ Ⓒ Ⓓ	84 Ⓐ Ⓑ Ⓒ Ⓓ	124 Ⓐ Ⓑ Ⓒ Ⓓ	164 Ⓐ Ⓑ Ⓒ Ⓓ
5 Ⓐ Ⓑ Ⓒ Ⓓ	45 Ⓐ Ⓑ Ⓒ Ⓓ	85 Ⓐ Ⓑ Ⓒ Ⓓ	125 Ⓐ Ⓑ Ⓒ Ⓓ	165 Ⓐ Ⓑ Ⓒ Ⓓ
6 Ⓐ Ⓑ Ⓒ Ⓓ	46 Ⓐ Ⓑ Ⓒ Ⓓ	86 Ⓐ Ⓑ Ⓒ Ⓓ	126 Ⓐ Ⓑ Ⓒ Ⓓ	166 Ⓐ Ⓑ Ⓒ Ⓓ
7 Ⓐ Ⓑ Ⓒ Ⓓ	47 Ⓐ Ⓑ Ⓒ Ⓓ	87 Ⓐ Ⓑ Ⓒ Ⓓ	127 Ⓐ Ⓑ Ⓒ Ⓓ	167 Ⓐ Ⓑ Ⓒ Ⓓ
8 Ⓐ Ⓑ Ⓒ Ⓓ	48 Ⓐ Ⓑ Ⓒ Ⓓ	88 Ⓐ Ⓑ Ⓒ Ⓓ	128 Ⓐ Ⓑ Ⓒ Ⓓ	168 Ⓐ Ⓑ Ⓒ Ⓓ
9 Ⓐ Ⓑ Ⓒ Ⓓ	49 Ⓐ Ⓑ Ⓒ Ⓓ	89 Ⓐ Ⓑ Ⓒ Ⓓ	129 Ⓐ Ⓑ Ⓒ Ⓓ	169 Ⓐ Ⓑ Ⓒ Ⓓ
10 Ⓐ Ⓑ Ⓒ Ⓓ	50 Ⓐ Ⓑ Ⓒ Ⓓ	90 Ⓐ Ⓑ Ⓒ Ⓓ	130 Ⓐ Ⓑ Ⓒ Ⓓ	170 Ⓐ Ⓑ Ⓒ Ⓓ
11 Ⓐ Ⓑ Ⓒ Ⓓ	51 Ⓐ Ⓑ Ⓒ Ⓓ	91 Ⓐ Ⓑ Ⓒ Ⓓ	131 Ⓐ Ⓑ Ⓒ Ⓓ	171 Ⓐ Ⓑ Ⓒ Ⓓ
12 Ⓐ Ⓑ Ⓒ Ⓓ	52 Ⓐ Ⓑ Ⓒ Ⓓ	92 Ⓐ Ⓑ Ⓒ Ⓓ	132 Ⓐ Ⓑ Ⓒ Ⓓ	172 Ⓐ Ⓑ Ⓒ Ⓓ
13 Ⓐ Ⓑ Ⓒ Ⓓ	53 Ⓐ Ⓑ Ⓒ Ⓓ	93 Ⓐ Ⓑ Ⓒ Ⓓ	133 Ⓐ Ⓑ Ⓒ Ⓓ	173 Ⓐ Ⓑ Ⓒ Ⓓ
14 Ⓐ Ⓑ Ⓒ Ⓓ	54 Ⓐ Ⓑ Ⓒ Ⓓ	94 Ⓐ Ⓑ Ⓒ Ⓓ	134 Ⓐ Ⓑ Ⓒ Ⓓ	174 Ⓐ Ⓑ Ⓒ Ⓓ
15 Ⓐ Ⓑ Ⓒ Ⓓ	55 Ⓐ Ⓑ Ⓒ Ⓓ	95 Ⓐ Ⓑ Ⓒ Ⓓ	135 Ⓐ Ⓑ Ⓒ Ⓓ	175 Ⓐ Ⓑ Ⓒ Ⓓ
16 Ⓐ Ⓑ Ⓒ Ⓓ	56 Ⓐ Ⓑ Ⓒ Ⓓ	96 Ⓐ Ⓑ Ⓒ Ⓓ	136 Ⓐ Ⓑ Ⓒ Ⓓ	176 Ⓐ Ⓑ Ⓒ Ⓓ
17 Ⓐ Ⓑ Ⓒ Ⓓ	57 Ⓐ Ⓑ Ⓒ Ⓓ	97 Ⓐ Ⓑ Ⓒ Ⓓ	137 Ⓐ Ⓑ Ⓒ Ⓓ	177 Ⓐ Ⓑ Ⓒ Ⓓ
18 Ⓐ Ⓑ Ⓒ Ⓓ	58 Ⓐ Ⓑ Ⓒ Ⓓ	98 Ⓐ Ⓑ Ⓒ Ⓓ	138 Ⓐ Ⓑ Ⓒ Ⓓ	178 Ⓐ Ⓑ Ⓒ Ⓓ
19 Ⓐ Ⓑ Ⓒ Ⓓ	59 Ⓐ Ⓑ Ⓒ Ⓓ	99 Ⓐ Ⓑ Ⓒ Ⓓ	139 Ⓐ Ⓑ Ⓒ Ⓓ	179 Ⓐ Ⓑ Ⓒ Ⓓ
20 Ⓐ Ⓑ Ⓒ Ⓓ	60 Ⓐ Ⓑ Ⓒ Ⓓ	100 Ⓐ Ⓑ Ⓒ Ⓓ	140 Ⓐ Ⓑ Ⓒ Ⓓ	180 Ⓐ Ⓑ Ⓒ Ⓓ
21 Ⓐ Ⓑ Ⓒ Ⓓ	61 Ⓐ Ⓑ Ⓒ Ⓓ	101 Ⓐ Ⓑ Ⓒ Ⓓ	141 Ⓐ Ⓑ Ⓒ Ⓓ	181 Ⓐ Ⓑ Ⓒ Ⓓ
22 Ⓐ Ⓑ Ⓒ Ⓓ	62 Ⓐ Ⓑ Ⓒ Ⓓ	102 Ⓐ Ⓑ Ⓒ Ⓓ	142 Ⓐ Ⓑ Ⓒ Ⓓ	182 Ⓐ Ⓑ Ⓒ Ⓓ
23 Ⓐ Ⓑ Ⓒ Ⓓ	63 Ⓐ Ⓑ Ⓒ Ⓓ	103 Ⓐ Ⓑ Ⓒ Ⓓ	143 Ⓐ Ⓑ Ⓒ Ⓓ	183 Ⓐ Ⓑ Ⓒ Ⓓ
24 Ⓐ Ⓑ Ⓒ Ⓓ	64 Ⓐ Ⓑ Ⓒ Ⓓ	104 Ⓐ Ⓑ Ⓒ Ⓓ	144 Ⓐ Ⓑ Ⓒ Ⓓ	184 Ⓐ Ⓑ Ⓒ Ⓓ
25 Ⓐ Ⓑ Ⓒ Ⓓ	65 Ⓐ Ⓑ Ⓒ Ⓓ	105 Ⓐ Ⓑ Ⓒ Ⓓ	145 Ⓐ Ⓑ Ⓒ Ⓓ	185 Ⓐ Ⓑ Ⓒ Ⓓ
26 Ⓐ Ⓑ Ⓒ Ⓓ	66 Ⓐ Ⓑ Ⓒ Ⓓ	106 Ⓐ Ⓑ Ⓒ Ⓓ	146 Ⓐ Ⓑ Ⓒ Ⓓ	186 Ⓐ Ⓑ Ⓒ Ⓓ
27 Ⓐ Ⓑ Ⓒ Ⓓ	67 Ⓐ Ⓑ Ⓒ Ⓓ	107 Ⓐ Ⓑ Ⓒ Ⓓ	147 Ⓐ Ⓑ Ⓒ Ⓓ	187 Ⓐ Ⓑ Ⓒ Ⓓ
28 Ⓐ Ⓑ Ⓒ Ⓓ	68 Ⓐ Ⓑ Ⓒ Ⓓ	108 Ⓐ Ⓑ Ⓒ Ⓓ	148 Ⓐ Ⓑ Ⓒ Ⓓ	188 Ⓐ Ⓑ Ⓒ Ⓓ
29 Ⓐ Ⓑ Ⓒ Ⓓ	69 Ⓐ Ⓑ Ⓒ Ⓓ	109 Ⓐ Ⓑ Ⓒ Ⓓ	149 Ⓐ Ⓑ Ⓒ Ⓓ	189 Ⓐ Ⓑ Ⓒ Ⓓ
30 Ⓐ Ⓑ Ⓒ Ⓓ	70 Ⓐ Ⓑ Ⓒ Ⓓ	110 Ⓐ Ⓑ Ⓒ Ⓓ	150 Ⓐ Ⓑ Ⓒ Ⓓ	190 Ⓐ Ⓑ Ⓒ Ⓓ
31 Ⓐ Ⓑ Ⓒ Ⓓ	71 Ⓐ Ⓑ Ⓒ Ⓓ	111 Ⓐ Ⓑ Ⓒ Ⓓ	151 Ⓐ Ⓑ Ⓒ Ⓓ	191 Ⓐ Ⓑ Ⓒ Ⓓ
32 Ⓐ Ⓑ Ⓒ Ⓓ	72 Ⓐ Ⓑ Ⓒ Ⓓ	112 Ⓐ Ⓑ Ⓒ Ⓓ	152 Ⓐ Ⓑ Ⓒ Ⓓ	192 Ⓐ Ⓑ Ⓒ Ⓓ
33 Ⓐ Ⓑ Ⓒ Ⓓ	73 Ⓐ Ⓑ Ⓒ Ⓓ	113 Ⓐ Ⓑ Ⓒ Ⓓ	153 Ⓐ Ⓑ Ⓒ Ⓓ	193 Ⓐ Ⓑ Ⓒ Ⓓ
34 Ⓐ Ⓑ Ⓒ Ⓓ	74 Ⓐ Ⓑ Ⓒ Ⓓ	114 Ⓐ Ⓑ Ⓒ Ⓓ	154 Ⓐ Ⓑ Ⓒ Ⓓ	194 Ⓐ Ⓑ Ⓒ Ⓓ
35 Ⓐ Ⓑ Ⓒ Ⓓ	75 Ⓐ Ⓑ Ⓒ Ⓓ	115 Ⓐ Ⓑ Ⓒ Ⓓ	155 Ⓐ Ⓑ Ⓒ Ⓓ	195 Ⓐ Ⓑ Ⓒ Ⓓ
36 Ⓐ Ⓑ Ⓒ Ⓓ	76 Ⓐ Ⓑ Ⓒ Ⓓ	116 Ⓐ Ⓑ Ⓒ Ⓓ	156 Ⓐ Ⓑ Ⓒ Ⓓ	196 Ⓐ Ⓑ Ⓒ Ⓓ
37 Ⓐ Ⓑ Ⓒ Ⓓ	77 Ⓐ Ⓑ Ⓒ Ⓓ	117 Ⓐ Ⓑ Ⓒ Ⓓ	157 Ⓐ Ⓑ Ⓒ Ⓓ	197 Ⓐ Ⓑ Ⓒ Ⓓ
38 Ⓐ Ⓑ Ⓒ Ⓓ	78 Ⓐ Ⓑ Ⓒ Ⓓ	118 Ⓐ Ⓑ Ⓒ Ⓓ	158 Ⓐ Ⓑ Ⓒ Ⓓ	198 Ⓐ Ⓑ Ⓒ Ⓓ
39 Ⓐ Ⓑ Ⓒ Ⓓ	79 Ⓐ Ⓑ Ⓒ Ⓓ	119 Ⓐ Ⓑ Ⓒ Ⓓ	159 Ⓐ Ⓑ Ⓒ Ⓓ	199 Ⓐ Ⓑ Ⓒ Ⓓ
40 Ⓐ Ⓑ Ⓒ Ⓓ	80 Ⓐ Ⓑ Ⓒ Ⓓ	120 Ⓐ Ⓑ Ⓒ Ⓓ	160 Ⓐ Ⓑ Ⓒ Ⓓ	200 Ⓐ Ⓑ Ⓒ Ⓓ

TEST BOOKLET NUMBER

ANSWER SHEET

Directions for Marking the Answer Sheet

- Use a No. 2 lead pencil. Do **NOT** use ink or ball point pen.
- Make dark marks that completely fill the circle.
- Make **NO** stray marks on the answer sheet.

A NAME (Please print clearly.)

Last Name   First Name   Middle Initial

B FIRST 4 LETTERS OF LAST NAME

C APPLICANT ID NUMBER

D BIRTHDAY

MONTH | DAY

Jan. | Feb. | Mar. | Apr. | May | June | July | Aug. | Sept. | Oct. | Nov. | Dec.

E HAVE YOU EVER TAKEN THIS EXAM BEFORE?

Yes   No

F TEST FORM NUMBER

G TEST CENTER NUMBER

ANSWER SHEET

TEST BOOKLET NUMBER

Directions for Marking the Answer Sheet

- Use a No. 2 lead pencil. Do **NOT** use ink or ball point pen.
- Make dark marks that completely fill the circle.
- Make **NO** stray marks on the answer sheet.

A **NAME** (Please print clearly.)

Last Name   First Name   Middle Initial

B FIRST 4 LETTERS OF LAST NAME

C APPLICANT ID NUMBER

D BIRTHDAY

MONTH   DAY

Jan.
Feb.
Mar.
Apr.
May
June
July
Aug.
Sept.
Oct.
Nov.
Dec.

E HAVE YOU EVER TAKEN THIS EXAM BEFORE?

Yes No

F TEST FORM NUMBER

G TEST CENTER NUMBER

APPENDIX C

SAMPLE TAX TABLES

SINGLE Persons—WEEKLY Payroll Period

(For Wages Paid in 2006)

If the wages are—		And the number of withholding allowances claimed is—										
At least	But less than	0	1	2	3	4	5	6	7	8	9	10
		The amount of income tax to be withheld is—										
$0	$55	$0	$0	$0	$0	$0	$0	$0	$0	$0	$0	$0
55	60	1	0	0	0	0	0	0	0	0	0	0
60	65	1	0	0	0	0	0	0	0	0	0	0
65	70	2	0	0	0	0	0	0	0	0	0	0
70	75	2	0	0	0	0	0	0	0	0	0	0
75	80	3	0	0	0	0	0	0	0	0	0	0
80	85	3	0	0	0	0	0	0	0	0	0	0
85	90	4	0	0	0	0	0	0	0	0	0	0
90	95	4	0	0	0	0	0	0	0	0	0	0
95	100	5	0	0	0	0	0	0	0	0	0	0
100	105	5	0	0	0	0	0	0	0	0	0	0
105	110	6	0	0	0	0	0	0	0	0	0	0
110	115	6	0	0	0	0	0	0	0	0	0	0
115	120	7	0	0	0	0	0	0	0	0	0	0
120	125	7	1	0	0	0	0	0	0	0	0	0
125	130	8	1	0	0	0	0	0	0	0	0	0
130	135	8	2	0	0	0	0	0	0	0	0	0
135	140	9	2	0	0	0	0	0	0	0	0	0
140	145	9	3	0	0	0	0	0	0	0	0	0
145	150	10	3	0	0	0	0	0	0	0	0	0
150	155	10	4	0	0	0	0	0	0	0	0	0
155	160	11	4	0	0	0	0	0	0	0	0	0
160	165	11	5	0	0	0	0	0	0	0	0	0
165	170	12	5	0	0	0	0	0	0	0	0	0
170	175	12	6	0	0	0	0	0	0	0	0	0
175	180	13	6	0	0	0	0	0	0	0	0	0
180	185	13	7	0	0	0	0	0	0	0	0	0
185	190	14	7	1	0	0	0	0	0	0	0	0
190	195	14	8	1	0	0	0	0	0	0	0	0
195	200	15	8	2	0	0	0	0	0	0	0	0
200	210	16	9	3	0	0	0	0	0	0	0	0
210	220	18	10	4	0	0	0	0	0	0	0	0
220	230	19	11	5	0	0	0	0	0	0	0	0
230	240	21	12	6	0	0	0	0	0	0	0	0
240	250	22	13	7	0	0	0	0	0	0	0	0
250	260	24	14	8	1	0	0	0	0	0	0	0
260	270	25	16	9	2	0	0	0	0	0	0	0
270	280	27	17	10	3	0	0	0	0	0	0	0
280	290	28	19	11	4	0	0	0	0	0	0	0
290	300	30	20	12	5	0	0	0	0	0	0	0
300	310	31	22	13	6	0	0	0	0	0	0	0
310	320	33	23	14	7	1	0	0	0	0	0	0
320	330	34	25	15	8	2	0	0	0	0	0	0
330	340	36	26	17	9	3	0	0	0	0	0	0
340	350	37	28	18	10	4	0	0	0	0	0	0
350	360	39	29	20	11	5	0	0	0	0	0	0
360	370	40	31	21	12	6	0	0	0	0	0	0
370	380	42	32	23	13	7	1	0	0	0	0	0
380	390	43	34	24	14	8	2	0	0	0	0	0
390	400	45	35	26	16	9	3	0	0	0	0	0
400	410	46	37	27	17	10	4	0	0	0	0	0
410	420	48	38	29	19	11	5	0	0	0	0	0
420	430	49	40	30	20	12	6	0	0	0	0	0
430	440	51	41	32	22	13	7	0	0	0	0	0
440	450	52	43	33	23	14	8	1	0	0	0	0
450	460	54	44	35	25	15	9	2	0	0	0	0
460	470	55	46	36	26	17	10	3	0	0	0	0
470	480	57	47	38	28	18	11	4	0	0	0	0
480	490	58	49	39	29	20	12	5	0	0	0	0
490	500	60	50	41	31	21	13	6	0	0	0	0
500	510	61	52	42	32	23	14	7	1	0	0	0
510	520	63	53	44	34	24	15	8	2	0	0	0
520	530	64	55	45	35	26	16	9	3	0	0	0
530	540	66	56	47	37	27	18	10	4	0	0	0
540	550	67	58	48	38	29	19	11	5	0	0	0
550	560	69	59	50	40	30	21	12	6	0	0	0
560	570	70	61	51	41	32	22	13	7	1	0	0
570	580	72	62	53	43	33	24	14	8	2	0	0
580	590	73	64	54	44	35	25	16	9	3	0	0
590	600	75	65	56	46	36	27	17	10	4	0	0

SINGLE Persons—WEEKLY Payroll Period

(For Wages Paid in 2006)

If the wages are—		And the number of withholding allowances claimed is—										
At least	But less than	0	1	2	3	4	5	6	7	8	9	10
		The amount of income tax to be withheld is—										
$600	$610	$76	$67	$57	$47	$38	$28	$19	$11	$5	$0	$0
610	620	78	68	59	49	39	30	20	12	6	0	0
620	630	80	70	60	50	41	31	22	13	7	0	0
630	640	82	71	62	52	42	33	23	14	8	1	0
640	650	85	73	63	53	44	34	25	15	9	2	0
650	660	87	74	65	55	45	36	26	17	10	3	0
660	670	90	76	66	56	47	37	28	18	11	4	0
670	680	92	77	68	58	48	39	29	20	12	5	0
680	690	95	79	69	59	50	40	31	21	13	6	0
690	700	97	81	71	61	51	42	32	23	14	7	1
700	710	100	84	72	62	53	43	34	24	15	8	2
710	720	102	86	74	64	54	45	35	26	16	9	3
720	730	105	89	75	65	56	46	37	27	18	10	4
730	740	107	91	77	67	57	48	38	29	19	11	5
740	750	110	94	78	68	59	49	40	30	21	12	6
750	760	112	96	80	70	60	51	41	32	22	13	7
760	770	115	99	83	71	62	52	43	33	24	14	8
770	780	117	101	85	73	63	54	44	35	25	16	9
780	790	120	104	88	74	65	55	46	36	27	17	10
790	800	122	106	90	76	66	57	47	38	28	19	11
800	810	125	109	93	77	68	58	49	39	30	20	12
810	820	127	111	95	79	69	60	50	41	31	22	13
820	830	130	114	98	82	71	61	52	42	33	23	14
830	840	132	116	100	84	72	63	53	44	34	25	15
840	850	135	119	103	87	74	64	55	45	36	26	17
850	860	137	121	105	89	75	66	56	47	37	28	18
860	870	140	124	108	92	77	67	58	48	39	29	20
870	880	142	126	110	94	79	69	59	50	40	31	21
880	890	145	129	113	97	81	70	61	51	42	32	23
890	900	147	131	115	99	84	72	62	53	43	34	24
900	910	150	134	118	102	86	73	64	54	45	35	26
910	920	152	136	120	104	89	75	65	56	46	37	27
920	930	155	139	123	107	91	76	67	57	48	38	29
930	940	157	141	125	109	94	78	68	59	49	40	30
940	950	160	144	128	112	96	80	70	60	51	41	32
950	960	162	146	130	114	99	83	71	62	52	43	33
960	970	165	149	133	117	101	85	73	63	54	44	35
970	980	167	151	135	119	104	88	74	65	55	46	36
980	990	170	154	138	122	106	90	76	66	57	47	38
990	1,000	172	156	140	124	109	93	77	68	58	49	39
1,000	1,010	175	159	143	127	111	95	79	69	60	50	41
1,010	1,020	177	161	145	129	114	98	82	71	61	52	42
1,020	1,030	180	164	148	132	116	100	84	72	63	53	44
1,030	1,040	182	166	150	134	119	103	87	74	64	55	45
1,040	1,050	185	169	153	137	121	105	89	75	66	56	47
1,050	1,060	187	171	155	139	124	108	92	77	67	58	48
1,060	1,070	190	174	158	142	126	110	94	78	69	59	50
1,070	1,080	192	176	160	144	129	113	97	81	70	61	51
1,080	1,090	195	179	163	147	131	115	99	83	72	62	53
1,090	1,100	197	181	165	149	134	118	102	86	73	64	54
1,100	1,110	200	184	168	152	136	120	104	88	75	65	56
1,110	1,120	202	186	170	154	139	123	107	91	76	67	57
1,120	1,130	205	189	173	157	141	125	109	93	78	68	59
1,130	1,140	207	191	175	159	144	128	112	96	80	70	60
1,140	1,150	210	194	178	162	146	130	114	98	83	71	62
1,150	1,160	212	196	180	164	149	133	117	101	85	73	63
1,160	1,170	215	199	183	167	151	135	119	103	88	74	65
1,170	1,180	217	201	185	169	154	138	122	106	90	76	66
1,180	1,190	220	204	188	172	156	140	124	108	93	77	68
1,190	1,200	222	206	190	174	159	143	127	111	95	79	69
1,200	1,210	225	209	193	177	161	145	129	113	98	82	71
1,210	1,220	227	211	195	179	164	148	132	116	100	84	72
1,220	1,230	230	214	198	182	166	150	134	118	103	87	74
1,230	1,240	232	216	200	184	169	153	137	121	105	89	75
1,240	1,250	235	219	203	187	171	155	139	123	108	92	77

$1,250 and over   Use Table 1(a) for a **SINGLE person** on page 36. Also see the instructions on page 34.

MARRIED Persons—WEEKLY Payroll Period

(For Wages Paid in 2006)

If the wages are—		And the number of withholding allowances claimed is—										
At least	But less than	0	1	2	3	4	5	6	7	8	9	10
		The amount of income tax to be withheld is—										
$0	$125	$0	$0	$0	$0	$0	$0	$0	$0	$0	$0	$0
125	130	0	0	0	0	0	0	0	0	0	0	0
130	135	0	0	0	0	0	0	0	0	0	0	0
135	140	0	0	0	0	0	0	0	0	0	0	0
140	145	0	0	0	0	0	0	0	0	0	0	0
145	150	0	0	0	0	0	0	0	0	0	0	0
150	155	0	0	0	0	0	0	0	0	0	0	0
155	160	0	0	0	0	0	0	0	0	0	0	0
160	165	1	0	0	0	0	0	0	0	0	0	0
165	170	1	0	0	0	0	0	0	0	0	0	0
170	175	2	0	0	0	0	0	0	0	0	0	0
175	180	2	0	0	0	0	0	0	0	0	0	0
180	185	3	0	0	0	0	0	0	0	0	0	0
185	190	3	0	0	0	0	0	0	0	0	0	0
190	195	4	0	0	0	0	0	0	0	0	0	0
195	200	4	0	0	0	0	0	0	0	0	0	0
200	210	5	0	0	0	0	0	0	0	0	0	0
210	220	6	0	0	0	0	0	0	0	0	0	0
220	230	7	1	0	0	0	0	0	0	0	0	0
230	240	8	2	0	0	0	0	0	0	0	0	0
240	250	9	3	0	0	0	0	0	0	0	0	0
250	260	10	4	0	0	0	0	0	0	0	0	0
260	270	11	5	0	0	0	0	0	0	0	0	0
270	280	12	6	0	0	0	0	0	0	0	0	0
280	290	13	7	0	0	0	0	0	0	0	0	0
290	300	14	8	1	0	0	0	0	0	0	0	0
300	310	15	9	2	0	0	0	0	0	0	0	0
310	320	16	10	3	0	0	0	0	0	0	0	0
320	330	17	11	4	0	0	0	0	0	0	0	0
330	340	18	12	5	0	0	0	0	0	0	0	0
340	350	19	13	6	0	0	0	0	0	0	0	0
350	360	20	14	7	1	0	0	0	0	0	0	0
360	370	21	15	8	2	0	0	0	0	0	0	0
370	380	22	16	9	3	0	0	0	0	0	0	0
380	390	23	17	10	4	0	0	0	0	0	0	0
390	400	24	18	11	5	0	0	0	0	0	0	0
400	410	25	19	12	6	0	0	0	0	0	0	0
410	420	26	20	13	7	1	0	0	0	0	0	0
420	430	27	21	14	8	2	0	0	0	0	0	0
430	440	28	22	15	9	3	0	0	0	0	0	0
440	450	29	23	16	10	4	0	0	0	0	0	0
450	460	31	24	17	11	5	0	0	0	0	0	0
460	470	32	25	18	12	6	0	0	0	0	0	0
470	480	34	26	19	13	7	0	0	0	0	0	0
480	490	35	27	20	14	8	1	0	0	0	0	0
490	500	37	28	21	15	9	2	0	0	0	0	0
500	510	38	29	22	16	10	3	0	0	0	0	0
510	520	40	30	23	17	11	4	0	0	0	0	0
520	530	41	32	24	18	12	5	0	0	0	0	0
530	540	43	33	25	19	13	6	0	0	0	0	0
540	550	44	35	26	20	14	7	1	0	0	0	0
550	560	46	36	27	21	15	8	2	0	0	0	0
560	570	47	38	28	22	16	9	3	0	0	0	0
570	580	49	39	30	23	17	10	4	0	0	0	0
580	590	50	41	31	24	18	11	5	0	0	0	0
590	600	52	42	33	25	19	12	6	0	0	0	0
600	610	53	44	34	26	20	13	7	1	0	0	0
610	620	55	45	36	27	21	14	8	2	0	0	0
620	630	56	47	37	28	22	15	9	3	0	0	0
630	640	58	48	39	29	23	16	10	4	0	0	0
640	650	59	50	40	31	24	17	11	5	0	0	0
650	660	61	51	42	32	25	18	12	6	0	0	0
660	670	62	53	43	34	26	19	13	7	0	0	0
670	680	64	54	45	35	27	20	14	8	1	0	0
680	690	65	56	46	37	28	21	15	9	2	0	0
690	700	67	57	48	38	29	22	16	10	3	0	0
700	710	68	59	49	40	30	23	17	11	4	0	0
710	720	70	60	51	41	32	24	18	12	5	0	0
720	730	71	62	52	43	33	25	19	13	6	0	0
730	740	73	63	54	44	35	26	20	14	7	1	0

MARRIED Persons—WEEKLY Payroll Period
(For Wages Paid in 2006)

If the wages are—		And the number of withholding allowances claimed is—										
At least	But less than	0	1	2	3	4	5	6	7	8	9	10
		The amount of income tax to be withheld is—										
$740	$750	$74	$65	$55	$46	$36	$27	$21	$15	$8	$2	$0
750	760	76	66	57	47	38	28	22	16	9	3	0
760	770	77	68	58	49	39	30	23	17	10	4	0
770	780	79	69	60	50	41	31	24	18	11	5	0
780	790	80	71	61	52	42	33	25	19	12	6	0
790	800	82	72	63	53	44	34	26	20	13	7	1
800	810	83	74	64	55	45	36	27	21	14	8	2
810	820	85	75	66	56	47	37	28	22	15	9	3
820	830	86	77	67	58	48	39	29	23	16	10	4
830	840	88	78	69	59	50	40	31	24	17	11	5
840	850	89	80	70	61	51	42	32	25	18	12	6
850	860	91	81	72	62	53	43	34	26	19	13	7
860	870	92	83	73	64	54	45	35	27	20	14	8
870	880	94	84	75	65	56	46	37	28	21	15	9
880	890	95	86	76	67	57	48	38	29	22	16	10
890	900	97	87	78	68	59	49	40	30	23	17	11
900	910	98	89	79	70	60	51	41	32	24	18	12
910	920	100	90	81	71	62	52	43	33	25	19	13
920	930	101	92	82	73	63	54	44	35	26	20	14
930	940	103	93	84	74	65	55	46	36	27	21	15
940	950	104	95	85	76	66	57	47	38	28	22	16
950	960	106	96	87	77	68	58	49	39	30	23	17
960	970	107	98	88	79	69	60	50	41	31	24	18
970	980	109	99	90	80	71	61	52	42	33	25	19
980	990	110	101	91	82	72	63	53	44	34	26	20
990	1,000	112	102	93	83	74	64	55	45	36	27	21
1,000	1,010	113	104	94	85	75	66	56	47	37	28	22
1,010	1,020	115	105	96	86	77	67	58	48	39	29	23
1,020	1,030	116	107	97	88	78	69	59	50	40	31	24
1,030	1,040	118	108	99	89	80	70	61	51	42	32	25
1,040	1,050	119	110	100	91	81	72	62	53	43	34	26
1,050	1,060	121	111	102	92	83	73	64	54	45	35	27
1,060	1,070	122	113	103	94	84	75	65	56	46	37	28
1,070	1,080	124	114	105	95	86	76	67	57	48	38	29
1,080	1,090	125	116	106	97	87	78	68	59	49	40	30
1,090	1,100	127	117	108	98	89	79	70	60	51	41	32
1,100	1,110	128	119	109	100	90	81	71	62	52	43	33
1,110	1,120	130	120	111	101	92	82	73	63	54	44	35
1,120	1,130	131	122	112	103	93	84	74	65	55	46	36
1,130	1,140	133	123	114	104	95	85	76	66	57	47	38
1,140	1,150	134	125	115	106	96	87	77	68	58	49	39
1,150	1,160	136	126	117	107	98	88	79	69	60	50	41
1,160	1,170	137	128	118	109	99	90	80	71	61	52	42
1,170	1,180	139	129	120	110	101	91	82	72	63	53	44
1,180	1,190	140	131	121	112	102	93	83	74	64	55	45
1,190	1,200	142	132	123	113	104	94	85	75	66	56	47
1,200	1,210	143	134	124	115	105	96	86	77	67	58	48
1,210	1,220	145	135	126	116	107	97	88	78	69	59	50
1,220	1,230	146	137	127	118	108	99	89	80	70	61	51
1,230	1,240	148	138	129	119	110	100	91	81	72	62	53
1,240	1,250	149	140	130	121	111	102	92	83	73	64	54
1,250	1,260	151	141	132	122	113	103	94	84	75	65	56
1,260	1,270	152	143	133	124	114	105	95	86	76	67	57
1,270	1,280	154	144	135	125	116	106	97	87	78	68	59
1,280	1,290	155	146	136	127	117	108	98	89	79	70	60
1,290	1,300	157	147	138	128	119	109	100	90	81	71	62
1,300	1,310	158	149	139	130	120	111	101	92	82	73	63
1,310	1,320	161	150	141	131	122	112	103	93	84	74	65
1,320	1,330	163	152	142	133	123	114	104	95	85	76	66
1,330	1,340	166	153	144	134	125	115	106	96	87	77	68
1,340	1,350	168	155	145	136	126	117	107	98	88	79	69
1,350	1,360	171	156	147	137	128	118	109	99	90	80	71
1,360	1,370	173	158	148	139	129	120	110	101	91	82	72
1,370	1,380	176	160	150	140	131	121	112	102	93	83	73
1,380	1,390	178	162	151	142	132	123	113	104	94	85	75
1,390	1,400	181	165	153	143	134	124	115	105	96	86	

$1,400 and over   Use Table 1(b) for a **MARRIED person** on page 36. Also see the instructions on page 34.

APPENDIX D

STATE ELECTRICAL CONTRACTING WEB SITES

Contractor's License Reference Site
http://www.contractors-license.org/

Alabama
www.aecb.state.al.us

Alaska
http://www.commerce.state.ak.us/
choose Division of Corporations, Business, and Professional Licensing under Professions Licensed, choose Electrical Administrators

Arizona
http://www.azroc.gov/
on left menu highlight Licensing > choose License Classifications

Arkansas
http://www.state.ar.us/
search for Contractors Licensing Board > choose Arkansas State Contractors Board

California
http://www.cslb.ca.gov/

Colorado
http://www.dora.state.co.us/
on left menu choose Division of Registrations > on left menu under Licensing Boards and Programs choose Electricians

Connecticut
http://www.ct.gov/
in the Quick Links drop-down menu choose State Agencies and press Go > find Consumer Protection, Department of

Delaware
http://www.dpr.delaware.gov
choose Electrician from the list of professions

Florida
http://www.myflorida.com
at the top of the page choose Find an Agency > choose Business & Professional Regulation > choose Electrical Contractors Licensing

Georgia
http://www.sos.state.ga.us/
choose Professional Licensure

Hawaii
http://www.hawaii.gov/
choose the Go button for Government > under State Government choose Department of Commerce and Consumer Affairs > choose Professional and Vocational Licensing (PVL)

Idaho
http://dbs.idaho.gov/
on left menu choose Electrical Bureau

Illinois
At the time of publishing, no state exam required. See local county or city requirements for electrical contracting information.

Indiana
At the time of publishing, no state exam required. See local county or city requirements for electrical contracting information.

Iowa
http://www.iowaworkforce.org/
on left menu highlight Labor Services Division > choose Contractor Registration

Kansas
At the time of publishing, no state exam required. See local county or city requirements for electrical contracting information.

Kentucky
http://kentucky.gov
from the top menu choose KY Agencies > under H choose Housing, Buildings and Construction, Office of > from the left menu choose Licensing, Plan Reviews, Inspections and Fees > from the left menu choose Electrical

Louisiana
http://www.lslbc.state.la.us/

Maine
http://www.state.me.us/
from the top menu choose State Agencies > under P choose *Professional and Financial Regulation, Dept. of* from the menu choose Professional Licensing

lr.state.md.us/
nal and Professional Licensing choose Licensing Boards

State Agencies > choose Alphabetic List, All Branches > under P choose Professional
ss button to Enter > choose Boards of Registration > under E choose Electricians and

and Agencies > scroll down to choose Labor and Economic Growth > from
from the left menu choose Examinations and Licensing

Mississippi
http://www.msboc.us/

Missouri
At the time of publishing, no state exam required. See local county or city requirements for electrical contracting information.

Montana
http://mt.gov/
from the left menu choose State Agencies > under Labor and Industry Department choose Business Standards Division > in the drop-down menu choose Electrician

Nebraska
http://www.electrical.state.ne.us/

Nevada
http://www.nvcontractorsboard.com/
choose the link for Contractor Information

New Hampshire
http://www.nh.gov/
under State Government choose State Agencies > under Safety Department choose Electrical Safety and Licensing Bureau

New Jersey
http://www.state.nj.us/
at the top choose Departments/Agencies > choose Attorney General, Office of the (Law & Public Safety) > from the left menu choose Division of Consumer Affairs > in the green bar choose Licensing Boards > choose Other Professional and Occupational Licensing Boards > choose Electrical Contractors, Board of Examiners of

New Mexico
http://www.rld.state.nm.us/
from the left menu choose Regulation and Licensing Department > choose Construction Industries

New York
At the time of publishing, no state exam required. See local county or city requirements for electrical contracting information.

North Carolina
http://www.ncbeec.org/

North Dakota
http://www.ndseb.com/

Ohio
http://www.com.state.oh.us/
at the top choose eLicense Center > from the left menu choose Industrial Compliance > choose Ohio Constructi Industry Licensing Board

Oklahoma
http://www.cib.state.ok.us/

Oregon
http://www.ccb.state.or.us/

Pennsylvania
At the time of publishing, no state exam required. See local county or city requirements for electrical contracting information.

Rhode Island
http://www.crb.ri.gov/

South Carolina
http://www.llr.state.sc.us/
from the left menu choose Licensing Boards > from the top pull-down menu choose Contractors

South Dakota
http://www.state.sd.us/
highlight Government > highlight Agencies > highlight Labor > from the left menu under Resources choose Boards and Commissions > choose South Dakota Electrical Commission

Tennessee
http://www.state.tn.us/
choose Government > under State choose List of Departments and Agencies > choose Department of Commerce & Insurance > from the left menu choose Regulatory Boards > scroll down to choose Contractors/Home Improvement

Texas
http://www.license.state.tx.us/
in the pull-down menu choose Electricians

Utah
http://www.dopl.utah.gov/
choose Licensing > Choose Occupations & Professions > choose Electrician

Vermont
http://www.dps.state.vt.us/
click the link for Division of Fire Safety > in the left menu under Licensing/Certification choose Electrical

Virginia
http://www.dpor.virginia.gov/
the left menu choose Licensing and Regulations (Boards) > scroll down to choose Contractors under Boards
ites

n

i.wa.gov/
e Trades & Licensing

hal.org/
down to choose Electricians

on

/
afety and Buildings Division

OTHER HELPFUL WEB SITES
FOR INFORMATION AND BOOKS

http://www.appliedlearningsol.com
Electrical videos and exam information

http://www.nfpacatalog.com
National Fire Protection Association

http://www.delmarlearning.com
Textbooks and videos

http://www.irs.gov
Internal Revenue Service

http://www.nascla.org
National Association of State Contractors Licensing Agencies
Books for contractors such as business and management

For more resources concerning electrical contractor's license review, please contact Applied Learning Solutions, LLC
Web: *www.electricalexamsolutions.com*
Email: *examsolutions@comcast.net*

Note: Page numbers followed by a "t" or "f" indicate that the entry is included in a table or figure. Article and table numbers in parenthesis correspond to the National Electrical Code.